The Curse of American Agricultural Abundance

VOLUME 16 IN THE SERIES
Our Sustainable Future

Series Editors

Charles A. Francis
University of Nebraska–Lincoln

Cornelia Flora
Iowa State University

Paul A. Olson
University of Nebraska–Lincoln

The Curse *of* American Agricultural Abundance

A Sustainable Solution

Willard W. Cochrane

WITH A FOREWORD BY RICHARD A. LEVINS

University of Nebraska Press | Lincoln and London

Chapter 1 was published as "The Case for Production Control" in *Metropolitan Milk Producer News* 14, no. 12 (Dec. 1954), printed by the Metropolitan Cooperative Milk Producers' Bargaining Agency, Inc., in Syracuse, New York.

Chapter 2 was published as chapter 5, "The Agricultural Treadmill," in *Farm Prices: Myth and Reality* (Minneapolis: University of Minnesota Press, 1958). © Copyright 1958 by the University of Minnesota.

Chapter 3 was published as "Farm Technology, Foreign Surplus Disposal and Domestic Supply Control" in *Journal of Farm Economics* 61, no. 5 (Dec. 1959).

Chapter 4 was published in summary form as "A Food and Agricultural Policy for the 21st Century" by the Institute for Agriculture and Trade Policy in Minneapolis, Minnesota.

"What Makes Sustainable Farms Successful" is reprinted from Elizabeth Bird, Gordon Bultena, and John Gardner, eds., *Planting the Future* (Ames: Iowa State University Press, 1995), pp. 123–37, with the permission of the Center for Rural Affairs, Walthill, Nebraska; research and publication underwritten by Northwest Area Foundation, St. Paul MN, publication coordinated by Center for Rural Affairs, Walthill NE.

Library of Congress Cataloging-in-
Publication Data
Cochrane, Willard Wesley, 1914–
The curse of
American agricultural abundance:
a sustainable solution /
Willard W. Cochrane;
with a foreword by Richard A. Levins.
p. cm. — (Our sustainable future : v. 16)
Includes bibliographical references
and index.
ISBN 0-8032-1529-0 (cloth : alk. paper)
1. Agriculture and state—United States.
2. Agriculture—Economic aspects—
United States.
3. Surplus agricultural commodities,
American.
I. Levins, Richard A.
II. Title.
III. Series.
HD 1761.C596 2003
338.1′873—dc21
2003040995

For E. C. Voorhies,
who got me started in this business in the 1930s,
and for John D. Black,
who provided guidance and support to
this young professional when it counted, in the 1940s.

CONTENTS

Foreword   ix

Acknowledgments   xi

Prologue: Who and Why?   1

PART 1. POLICIES OF THE MID-1900S

1. The Case for Production Control   9

2. The Agricultural Treadmill   19

3. Farm Technology, Foreign Surplus Disposal,
   and Domestic Supply Control   45

PART 2. POLICIES FOR 2002 AND BEYOND

4. A Food and Agricultural Policy
   for the Twenty-first Century   65

5. The Export Solution   83

6. Saving the Family Farm:
   The Case for Government Intervention   91

7. American Agricultural Abundance:
   Curse or Opportunity?   107

   Epilogue: The Future?   131

   Appendix: What Makes
   Sustainable Farms Successful?   135

   Notes   143

   Index   149

When Willard Cochrane went to Washington to serve as chief agricultural economist with the Kennedy administration, he was already one of the nation's premier academic policy economists. He knew all too well the statistics on how surplus farm production had become a national problem. Even so, in the many years I have had the pleasure to be his friend, he has never mentioned those statistics. He prefers to tell of his horror upon first learning that the program he inherited from the Eisenhower administration was renting several abandoned movie theaters out west to store surplus wheat.

It is one thing to see statistics on how many bears are in the woods, and quite another to find one in your tent. The movie theater story, and many more like it, remind us that Willard Cochrane saw the bear of surplus production in his tent. It became a beast to be tamed at all costs. Those costs were very high, both for the nation and for Cochrane personally. Farmers rejected the Kennedy program of supply control in the 1963 wheat referendum. The political backlash gave birth to the "no holds barred, let the government clean up the mess" style of farm policy that we see today. Cochrane blamed himself for the program failure and returned to his University of Minnesota classroom in 1964 both sadder and wiser.

Nonetheless, he never gave up on his search for solutions. Because of his brilliance and his depth of experience, he found little comfort in the ideas most others found appealing. For example, he saw no panacea in the old liberal standby of giving surplus food to the world's starving nations. When Cochrane advised the government of

India in post-Kennedy years, he watched with pride as ships filled with U.S. grain lined up in Calcutta's harbor. But he also knew that in the long run the poor were hungry because they were poor, not because of a world food shortage. Solving the problems of world poverty required development programs that well could be hindered by lasting dependence on imported foods.

To fully appreciate this book, one must understand that although Cochrane has the mind of an economist, he has the heart of an environmentalist. Rachel Carson published *Silent Spring* while he was in Washington, and Cochrane still remembers the storm it created at the U.S. Department of Agriculture. Our food system, as productive as it was, would forever more have a contentious environmental dimension yet to be satisfactorily resolved in farm legislation. Those who know Cochrane well are hardly surprised by his many calls for a more sustainable agriculture. He loved the West, the great outdoors, long before he became an agricultural economist. Over lunch, he is as likely to talk about backpacking trips in the High Sierra or his latest reading of *Lonesome Dove* as the topic of farm policy.

It takes a lifetime, and a very productive and thoughtful one at that, to write a book like this one. As the years rolled along, Cochrane remained keenly interested in farm policy but became increasingly disenchanted with traditional farm programs. He gradually came to see the environment, food safety, the increasing power of agribusiness, and globalization as issues beyond anything conventional farm programs were designed to handle. In *The Curse of American Agricultural Abundance: A Sustainable Solution*, Cochrane outlines a vast transformation of the American food system. That transformation is one in which the curse of the past becomes the opportunity of the future.

Richard A. Levins

ACKNOWLEDGMENTS

In this project, as in most serious writing projects, many people contributed to the formulation of the ideas that emerge. Teachers and colleagues, too many to list, had a hand in my intellectual development. But I will mention a few who had the most influence on my thinking about American farm policy.

Most of the people who influenced my thinking, and hence the writing of the three essays in part 1, are now dead. But they were so important in my early intellectual development that I wish to credit them here: Bushrod W. Allin, James G. Maddox, and John Brewster of the old Bureau of Agricultural Economics (BAE), Nathan Koffsky and Harry Trelogan of the U.S. Department of Agriculture, my co-author Walter W. Wilcox, and my friendly critic T. W. Schultz. Those that happily are still with us include George E. Brandow, Elmer Learn, Lee Day, and Sherwood O. Berg.

A good number of people contributed both directly and indirectly to the ideas that emerge in the four essays in part 2. Richard A. Levins encouraged me to undertake this project and critically reviewed each essay. My office mate, W. Burt Sundquist, critically reviewed most of the writing and made helpful suggestions. My modern-day friendly critic Vernon Ruttan encouraged me to give farm policy reform one more try.

Mark Ritchie, president of the Institute for Agriculture and Trade Policy, asked me to prepare the paper "A Food and Agriculture Policy for the 21st Century," included in this volume as chapter 4, for the use of his organization. Kathryn Gilje of that organization made some important suggestions for improving the paper, and Ben Lilliston, also of that organization, prepared short pieces for publication in midwestern newspapers from essays that appear as chapters 4 and 5 in this volume. And Jack Sperbeck of the University of Minnesota Extension Service also prepared short pieces for publication in various newspapers and magazines from the material in chapters 4 and 5.

Karl Stauber, president of the Northwest Area Foundation, was helpful by providing materials on sustainable farming that his organization had either developed or acquired. Various groups of active farmers to whom I presented ideas that became a part of this book helped me eliminate some of the bad ideas and sharpen the ones that remain.

Within the Department of Applied Economics at the University of Minnesota, two people were most helpful: Louise Letnes, our most efficient and gracious departmental librarian, helped me locate key references and data, and Jerry Fruin helped me untangle some difficult grain pricing concepts.

My secretary, Marilyn Clement, did wondrous work in transcribing my hand scribblings (which have become worse over eighty-eight years) into a readable typed text. And the Department of Applied Economics, under the leadership of Vernon Eidman, provided this professor emeritus with the friendly, solid support I needed to produce this volume.

My wife, Mary, supported and helped me, as always, in the development and editing of the manuscript.

PROLOGUE

Who and Why?

With brief exceptions, in one way or another I have been associated with agriculture all my life. As a boy, I lived on large ranches in California that my father managed. This ranch life was interspersed with summer visits to my grandparents' family farm in southwestern Iowa. I graduated from the University of California, Berkeley, in 1937 with a degree in agricultural economics. I then did graduate work in agricultural economics at Montana State University, the University of Minnesota, and Harvard University. I did four stints with the federal government dealing with agricultural matters: with the Cooperative Research and Service Division of the Farm Credit Administration in 1939–41; with the War Food Administration in 1943; with the old Bureau of Agricultural Economics (BAE) in 1944–47; and as the economic adviser to the secretary of agriculture in 1961–64. I did research and teaching in agricultural economics at Pennsylvania State University in 1949–51 and at the University of Minnesota from 1951 to 1981.

During the 1950s and 1960s, I was active in farm policy debates and in Democratic farm politics. I did staff work for Senator Hubert Humphrey, served as chairman of Governor Orville Freeman's Study Commission on Agriculture in 1957–58, and served on the personal staff of Senator John F. Kennedy during his run for the presidency in 1960, helping him formulate his food and agriculture policy. During my professional career, I traveled the world as an agricultural adviser, consultant, and mission member. My work ranged from taking part in the United Nations mission to Siam (now Thailand) in

1948 to serving as an agricultural adviser to the government of India, sponsored by the Ford Foundation (1965–75), to attending repeated meetings and conferences of the United Nations Food and Agriculture Organization in Rome (between 1960 and 1980), to assisting the government of Saudi Arabia (1973–74) with its planning efforts in agriculture.

I have observed firsthand the development of American agriculture in the twentieth century: the wondrous technological developments and increases in productivity; the dramatic decline in the number of family farms and the equally dramatic increase in the size of a relatively few large farms; and the rise of the great agribusiness firms providing new and improved inputs on one hand and advances in the handling and processing of farm products on the other.

Agriculture has moved, literally, from horse-drawn machines at the turn of the twentieth century to an industrialized state at the turn of the twenty-first, making use of the most advanced technologies. Neither my father nor my grandfather, both of whom left this vale of tears in the latter part of the 1920s, would recognize what is now taking place on the ranches and farms they once worked. They would be lost in this new technological world.

But two things about American agriculture seem never to change: first, its capacity to produce abundantly and, second, the inability of our social and economic system to make effective use of that abundance. The propensity for American agriculture to produce too much, and the failure of our social and economic system to make effective use of that unending abundance, is the subject of this book. We will look at two periods when abundant supplies, pressing on a limited commercial demand, resulted in low farm prices and incomes. The two periods are 1953–66 and 1997–2002 and beyond. Structurally, technologically, and organizationally the two periods are as different as night and day. But in terms of the big economic picture—too much production pressing on a limited commercial demand, resulting in low farm prices and incomes—they are mirror images.

We will explore the first period, 1953–66, through a speech I made to the Metropolitan Cooperative Milk Producers' Bargaining Agency of New York in 1954; one chapter from my book *Farm Prices: Myth and Reality* (1958); and my presidential address to the American Farm Economics Association in 1959. We will explore the second period, 1997–2002 and beyond, through four essays I have written since 1998: one for the Institute for Agricultural and Trade Policy of Minneapolis, Minnesota; one for a series on current farm problems for the University of Minnesota Extension Service; one for liberal farm organizations in the upper Midwest; and one, the final one, that I wrote specifically for this volume. What we have here is a book of essays covering two periods some fifty years apart.

These essays were not written in the order of chapters of a book. Each was meant for a specific audience at a specific time. Hence there is some duplication in the material, and perhaps some conflict of ideas. But all the essays from both periods deal with the same theme— the curse of too much production and what to do about it. Stated more formally, and perhaps more accurately, the American farm sector operates at full production all the time, and whenever the demand for its food products slackens or contracts, for whatever reason, the prices of those products fall, and fall sharply, because of the inelasticity of the demand, carrying farm incomes down with them. When these farm depressions occur, as they did numerous times in the 1900s (one of them lasted twenty years, 1921 to 1940), farmers have cried out to their government for help, and the government has on each occasion responded with some form of assistance. But with farmers, farm leaders, and politicians in disagreement over both the causes of farm depressions and solutions to them, the government's help has typically been ineffective, or at best a palliative. That is where we stand in the year 2002, in the midst of another farm depression.

But we need to recognize that the economic environment in which the whole economy operates, and particularly the farm economy, has undergone a profound change since the 1950s. Except for producers of cotton and wheat, which have been important American exports since colonial days, in the 1950s most farmers produced for the domestic market. By this I mean they produced commodities whose prices were determined by the forces of supply and demand in the United States. When farmers produced a surplus over domestic needs, as happened with broilers and soybeans and soybean meal in the 1950s, farmers and their agents sought specific foreign markets for those surplus products. American farmers in the 1950s were oriented to the domestic market and thought of foreign markets as a place to dispose of their surpluses, hopefully at good prices.

As of the early 2000s, not only producers of cotton and wheat, but producers of corn (and other feed grains), soybeans and soybean meal, rice, poultry products, pork products, and beef products know (or should know) that they are producing for a global market where the level of prices for each of these products is determined by the forces of supply and demand in that global market. This means that when policy solutions are being formulated to deal with depressed prices and incomes, American farmers must recognize that the low prices they are receiving were determined not in U.S. domestic markets but in international markets. When soybean producers in the United States give thought to policy changes with respect to their commodity, they must now also consider what the response will be in the soybean industries of Brazil and China. And when corn producers in the United States consider policy changes with respect to their commodity, they must also consider possible developments in the feed grain industries of Argentina, South Africa, and Eastern Europe. The global market for these commodities is one big, complex institutional arrangement in which producers in each exporting country can, and will, have an important impact on the producers in competing countries.

One important implication of the above is that a policy designed to reduce the production of, say, corn in the United States, in an effort to raise its price domestically, no longer makes economic sense. Cutting back on the production of corn in the United States in the context of a global market means first (assuming here that the U.S. action actually reduces the global supply of corn) that the United

States is bearing the full cost of raising corn prices for producers around the world. Second, it is improbable that reducing production in the United States would in fact reduce the total global supply of corn, because after a period of adjustment other producers around the world would increase production to supply any markets vacated by the United States' ill-advised decision to cut production unilaterally. The supply-control measures that I once advocated, and which were an integral part of American farm policy for over fifty years (albeit most often implemented ineffectively), no longer make sense for American farmers operating in a global market.

But this does not mean we cannot learn from the 1950s and 1960s. In my view the period 1997–2002 and beyond is a rerun of 1953–66 with respect to too much farm production, except that the market has expanded from a domestic one to a worldwide one. My three essays from the 1950s that make up part 1 of this book help us understand the powerful and dynamic role of technological advances in increasing farm production; help us understand how the domestic structure of farming, in combination with technological advances, operates to constantly push the national farm plant to higher and higher levels of production—to greater and greater aggregate output; and help us understand how difficult it is to turn this curse of too much production into opportunities that benefit both farmers and consumers.

Once again in the early 2000s, we are living with the curse of too much production and low farm prices and incomes. It is the purpose of this book to set forth a policy path that can turn this curse into an opportunity to create a healthy, sustainable farm economy. In chapters 4, 5, and 6 I analyze the problems confronting American farmers in the depressed period 1997–2002 and beyond and make some policy suggestions for dealing with those problems. In chapter 7 I develop those suggestions into a complete policy package that has, I believe, the power and capacity to create a farm economy in the United States that is both healthy and sustainable.

1 | Policies of the Mid-1900s

1 | THE CASE FOR PRODUCTION CONTROL

This essay is from a speech I gave to a group of dairy farmers in upstate New York in December 1954.[1] That was a period much like the present one, 1997–2002 and beyond, with dragging farm product prices, low family incomes, and much unhappiness in farming communities. There was no mystery, I argued, about the problem—farmers were producing too much. Technological developments following World War II had become a powerful force for increasing aggregate farm output. This rapidly increasing output, pressing against a slowly increasing demand, caused farm prices to fall; and prices fall sharply and dramatically because of the inelastic demand for food products. I concluded that we had to rely more heavily on production controls. Production controls would become the tools by which farmers, acting together with the support of government, could consciously and continuously adjust supplies to demand at prices deemed acceptable.

As I read the historical record of the United States, I see surplus stocks accumulating in agriculture and those stocks pressing against population needs in every peacetime period since the Civil War except 1900–1914. This pressure of food and fiber supplies against the needs of the population drives farm prices downward and creates the so-called farm problem. The present surplus condition is simply the current phase of a recurring pattern.

Bountiful supplies and low farm prices in the 1870s, 1880s, and 1890s grew out of an expansion of the national farm plant—additional workers on new lands—combined with some farm technological advances. The burdensome supplies of the 1920s and 1930s grew out of some acreage expansion during and immediately following World War I, the loss of foreign markets, and some important technological developments. The surpluses in 1948 and 1949 and from 1952 to date have grown almost entirely out of investments in new and expanded capital: improved breeding stock, tractors, irrigation systems, fertilizer, insect and disease control, and so on. We have witnessed and continue to witness a tremendous burst of technological advances on farms in the United States.

The period 1900–1914 is the one exception: during these years farm people enjoyed widespread prosperity. The question is, Why? The answer is that farm supplies ceased to press against demand. This was a time of extraordinary population increase: immigration reached an all-time high in this decade, and the rate of expansion in agricultural output slackened considerably. So in this peacetime period we have the demand for food products outpacing total supplies, causing farm prices and incomes to soar.

Output Exceeds Demand

We thus get the following picture: total farm output has increased and continues to increase year after year: once this expansion was based on new lands and increased workers; now it is based on farm technological advances. The total demand for the output of agriculture also has increased and continues to increase year after year; this expansion is based in part on population growth and in part on rising real incomes. *But in this race between expanding aggregate output and expanding aggregate demand, except in wartime and the one period noted, farm output has tended to push ahead of demand.* The result has been low prices and dragging incomes.

The question that immediately comes to mind is, Why in this race between output and demand does an imbalance between the two lead so readily to such dramatic price changes? How is it that we move so readily from a general condition of shortage to surplus, surplus to shortage, then back to surplus? For this is the record since 1946–47. The answer turns on the nature of the demand for food—on the regular demands of the human stomach. Most of us want to eat three times a day, and we eat about the same quantities of the same combinations of food every day. If the price of pork falls relative to the price of beef, we do substitute some pork for beef. And if the prices of all food products decline, we consume (eat or waste) a richer diet (note that I said a richer diet, not more pounds of food). But we do not stop eating when all food prices rise, and we do not greatly increase our consumption when all food prices fall. On the contrary, we go on eating about the same quantities of the same combinations of food.

Demand Is Inelastic

In the language of economists, we say that the demand for food is highly inelastic. More precisely, we know that the elasticity of aggregate demand for food approximates 0.3. Now, what does this mean? It means that consumers increase their consumption of all foods only 3 percent when the prices of all foods decline 10 percent. To farmers this means that *retail prices* must fall 10 percent to move into consumption a 3 percent increase in output. And where marketing margins absorb 50 percent of the consumer's food dollar, *farm prices* must fall in the neighborhood of 20 percent to move into consumption a 3 percent increase in output.

Now we begin to see one important source of the farm price problem. *At any given level of income and population size*, consumers want to go on eating about the same kinds and quantities of food. A cut of 3 percent in total food supplies means that farm prices will rise in the neighborhood of 20 percent; an increase of 3 percent in total food supplies leads to a decline in prices at the farm level of about 20 percent. The farm price consequences of a 6 percent increase or decrease in total food supplies are all too evident.

Since I am talking to milk producers, it is interesting to note that the elasticity of demand for fluid milk is about the same as that for all food—that is, 0.3. We are not so sure of this figure for fluid milk as we are for total food, but it represents our best estimate at present. So you producers find yourselves in a position like that of the average of all farmers—at the crack end of the whip. At any point in time, the prices producers receive for milk must fall a long way to induce people to step up their consumption by 3 percent.

There was a time when I believed the solution to these problems was to be found in increasing food consumption in the United States. And I still believe this is a good approach when people need the food. But the hard fact is that we are a well-fed nation. By any standard of comparison you may wish to use, the average consumer in the United States has a good diet.

A Look at Nutrition

But let us look more closely at the nutritional basis for increasing food consumption. Perhaps 25 percent of those families with incomes of less than $2,000 a year actually suffer from hunger pangs. And many more low-income families are underconsuming such nutrients as calcium and vitamin C. But on the average these same low-income families are consuming (eating or wasting) considerably more calories than good nutrition dictates. In the higher income brackets we find the same general picture, but less pronounced: the average consumer is overeating in terms of calories but failing to maintain an adequate intake of calcium and certain key vitamins. So we conclude that there are certain weak spots in American diets—particularly with respect to calcium. But in general consumers are overeating—they have too great a caloric intake. In terms of farm products this means that we need a general increase in the consumption of nonfat milk solids, possibly a greater intake of certain fruits and vegetables, but if anything a modest reduction in the overall poundage eaten.

But what of the argument that we could or should increase our consumption of animal proteins (red meat and eggs, for example)? From budget studies we know that high-income people eat more red meat than low-income people do. As we move up the income scale, the total number of pounds consumed increases very little, but the composition of the diet changes. Consumers shift away from cornmeal, wheat flour, and beans and toward red meats, poultry products, and fresh fruits and vegetables. But does this mean that people are voluntarily going to importantly increase their consumption of meat in the next few years? Perhaps, but I think not. Unless we get another spurt of economic activity such as we witnessed from 1950 to 1953, I believe the competition from other items in the budget—from new cars, home furnishings, and recreation—will if anything reduce the consumption of expensive meats and fresh fruits and vegetables. Remember, average consumers already eat well—each one consumed 154 pounds of meat in 1954; there is no great pressure on consumers to further enrich their diets.

Low-income consumers are another story. They would gladly buy more meat if they could afford to. But with their present incomes, they cannot. The question thus arises, Are we willing to subsidize low-income people's increased consumption of animal products? I for one would vote to do so, but to date Congress has shown no inclination to follow my lead.

The Low Price Approach

Another approach to the surplus problem in agriculture that we often hear talked about these days runs like this: lower the prices on those commodities in surplus to that point where they are no longer in surplus. Falling prices, it is argued, will reduce output as well as expand consumption and thus eliminate the surpluses. This approach, you will recognize, is currently being implemented in the form of lower levels of price support and flexible price supports. And some folks go even further; they argue that all forms of government support in agriculture, including marketing orders, should be tossed out the window.

Now, I did not come here to make a defense of price supports, but I do want to raise two questions with respect to this "get government out" approach. First, Will lower prices reduce farm output? Second, What are the consequences of this approach for farm people?

Most economists agree that the output of a commodity will decline in the short run in response to a decline in its price, if its producers have realistic production alternatives. For example, the reduction in the support price of wheat to 82.5 percent of parity will no doubt cause many producers in Michigan and Ohio to shift out of wheat and into some other commodity. But not so with wheat producers on the High Plains. Plains producers will stay in wheat because they do not have good alternatives. The whole question of production responses thus turns on the nature and number of alternative production opportunities.

Let the price of corn, and only corn, fall, and we can be sure that enough corn producers will shift into soybeans, wheat, cotton, and other commodities to reduce the supply of corn and place a floor under its falling price. But we cannot be so sure about wheat. As I mentioned, in the important wheat-producing areas of the High Plains farmers do not have good alternatives to raising wheat. Thus in the short run we should not expect falling wheat prices—even drastic declines—to reduce total wheat production importantly. Now, what may we expect from a decline in the prices of all those commodities that have come into surplus in the past two years—wheat, corn, vegetable oils, butter, nonfat milk solids, and cotton? We know first that individual producers are not, in the short run, going to cut their total production—total output per farm is going to be maintained. Producers will stay in the same enterprises if they have no good alternatives; or they may shift to a commodity already in trouble and further depress its price; or they may shift to a commodity not currently in trouble (for example, hogs) and pull down that price. The final result: the same aggregate production and lower prices all around. Falling prices in the short run will generally correct a surplus condition in a single commodity, but not a general surplus condition. In a general surplus situation, once again producers do not have good production alternatives.

I have been careful to say that falling prices all around will not reduce total farm output in the *short run*. And I don't think they will do so in the long run either. But lower farm prices and lower net incomes all around will do something. In the dynamic setting I discussed earlier, low farm prices and dragging incomes will slow down the rate of total output expansion. In the *long run*, then, we may expect this approach to solve the surplus problem by permitting demand to catch up with supply.

How does this come about, you ask? It comes about through the long and painful process of low prices, low incomes, financial distress, reduced levels of living, and business failure. In this context farmers cannot, or are unwilling to, invest in new and improved practices, new and more productive machinery and equipment, and improved facilities. Thus the rate of technological advance on farms is slowed down, and with it the rate of output expansion. The pressure of food supplies on demand is slowly relieved, and the surplus condition in agriculture is corrected.

This is the long, hard "let things alone" approach to the problem; it effects a solution through financial distress, personal suffering, and reduced technological advance. I know that I don't like this approach and I suspect that you don't. I don't like it for two principal reasons. First, a depressed agricultural sector acts as a drag or brake on the rest of the economy. I don't say that a depressed agriculture will necessarily pull the economy into a general business depression. But a depressed agriculture will pull down the level of national income unless that contraction is offset by an expansion in some other part of the economy.

Second, I would not like to see farm incomes decline further relative to urban incomes. My appraisal of the available income data, controversial as they may be, leads me to conclude that the median income (the exact middle income) in agriculture is considerably below the median urban income. This I think is undesirable, and I oppose any approach that, in my view, would widen this disparity.

What then is the answer to the basic problem of American agriculture? Trying to be realistic and fair, I don't think there is a single best solution. Some expansion in the food consumption of our low-income people is desirable, if it can be achieved. The expansion of foreign trade is desirable too, if it can be achieved. Both approaches certainly merit our serious consideration and support.

We Must Rely on Production Control

But I am not too hopeful with respect to either of these approaches. Further, and most important, neither can be turned on and off like a faucet whenever an imbalance develops between expanding supplies and expanding demand. The success of either depends on a continuing and sympathetic approach to the needs of the people involved. *Thus I conclude that we must rely more heavily on production control—* move further in the direction of consciously and continuously adjusting supplies to demand. The objective of this approach should be, it seems to me, an adjustment of supplies to demand that yields a pattern of farm prices that in turn produces incomes for farmers on a par with those of urban people. I am further convinced, as I think most people are, that the mechanics of control we now have at our disposal are inadequate and ineffective. (I refer to acreage controls, marketing quotas, and marketing orders and agreements.) Methods of control need to vary with the commodity, the area, and the problem. Let me illustrate by discussing four commodity problem areas and possible forms of production control. Please recognize that I realize these are not the last words to be said on production control. These are suggestions that I have picked up here and there, and in my view they need to be studied and possibly tried.

Specific Solutions

FLUID MILK MARKETS

Within the framework of federal order markets, it may become desirable to issue marketing quotas to each producer governing the amount of milk that each can sell for a fluid use price. The size of each producer's fluid quota would vary as the size of the total market quota varied, and the total market quota would be reviewed and determined periodically in accordance with the fluid milk needs of the market. Each producer would be paid the Class 1 price for milk sold under his quota; all other milk sold would go into manufacturing and

receive a manufacturing price. To introduce production flexibility—
to permit some producers to expand as others contract—these quotas
could be negotiable.

BUTTER PRODUCING AREAS IN NORTHEASTERN MINNESOTA

Here I would suggest developing programs designed to shift much of
the productive activities in this area from butterfat production to
livestock grazing. This would mean purchasing some farms and re-
grouping the remaining farms into units with larger acreage and
shifting the surplus population into nonfarm pursuits. Falling butterfat prices in this area will not effect this reorganization; falling prices
will more likely starve those involved into continued butterfat production. The surplus problem of this area requires conscious programming first to induce a production shift toward more extensive
uses, and second to help the displaced workers move into nonfarm
jobs.

WHEAT PRODUCTION IN THE HIGH-RISK AREAS
OF EASTERN COLORADO AND WESTERN KANSAS

Here I would suggest retiring this land from wheat production and
converting it back to grass. This would mean the development by the
federal government of a large-scale land purchase program and a
long-range grass restoration program. Once back in grass, the best
use of this land is obviously livestock grazing. It further seems wise to
hold this land in the public domain and lease its use to private interests under the federal grazing districts program.

CORN AND COTTON

Here I would experiment with a federal leasing scheme. In periods of
surplus stocks the federal government might rent corn and cotton
land and *hold it idle* in the optimum soil conserving practice of the
area—perhaps planted to a cover crop. In periods of shortages the
government would lower its leasing rates and permit the land to

move back into production. This scheme has some bugs in it, such as interfering with crop rotations and private leasing arrangements. But it has some real advantages: It is reversible—the government can get in and out easily. Reserves are held in the form of land rather than produced stocks. The leasing rates could be established on a net basis to cover only the fixed costs associated with the land. And most important, it solves the diverted acreage problem: the controlled acres under lease to the government could not be diverted into some other crop, hence serving as a conveyor belt for transferring the surplus condition from one commodity to another. Thus this approach seems to merit study and a trial.

Is Production Control Wrong?

Some will say that controlling production—that is, consciously and continuously adjusting supplies to demand—is socially and economically wrong. If this is true, then most modern business practices are wrong. Every large private business pursues the following general policy: first, create a market; second, stabilize that market; and finally supply that market in quantities that yield satisfactory incomes. And this, I suggest, is the direction that agriculture must and will go. But because many small units—family farms—are involved, those units must invoke the power of government to take concerted action.

2 | THE AGRICULTURAL TREADMILL

This essay is taken from my book Farm Prices: Myth and Reality, *published in 1958.[1] In it I explore the race between the aggregate demand for farm food products in the mid-1950s and the aggregate supply. I conclude that supply is outpacing demand and hence is acting to depress farm prices. I then advance my theory of why farmers keep adopting new and improved technologies and increasing output in the face of continuing low prices instead of reducing output as conventional thinking says they should. The theory explains why aggregate output keeps increasing in a depressed period, pressing on a restricted commercial demand for food, hence acting to prolong the farm depression. This theory of the agricultural treadmill remains as applicable an explanation of farmers' behavior in 1997–2002 and beyond as it was in the depressed 1950s.*

The Long-Run Race between
Aggregate Demand and Aggregate Supply

As we already know, the farm price level fluctuates in response to a shift in aggregate demand relative to supply, or a shift in aggregate supply relative to demand. But it is not correct to visualize these aggregate relations shifting back and forth in a static, no-growth context. Over the long run, both of these aggregate relations have been expanding: what we have had is a race between aggregate demand and aggregate supply. And changes in the farm price level that have occurred, growing out of shifts in the *relative* positions of the aggregate demand and supply relations, have most often resulted from unequal rates of expansion in these aggregate relations. The race has rarely been equal, and at times it has been very unequal, with extreme income consequences.[2]

During the nineteenth and early twentieth centuries, both rising real incomes and population growth operated to expand the aggregate demand for food. Rising real incomes enabled the average consumer to move away from a plain diet heavily weighted with potatoes and cereals to a varied and expensive diet—varied in terms of more animal products, more fruits and vegetables in and out of season, and more delicacies (cheeses, seafood, baked goods), and expensive in terms of greater dollar cost and more farm resources required to produce it. And population growth contributed more mouths to feed.

It is generally believed that the population elasticity for food approximates 1.0—meaning that a 1 percent increase in population results in a 1 percent increase in food consumption. This population elasticity estimate will vary as the means of population growth (immigration and natural increase) varies, but it is probably a useful rule of thumb. And since the total population of the United States increased by about 2,000 percent between 1800 and 1920, it follows that the aggregate demand for food increased by roughly the same amount as the result of population growth. During this long period, the market for farm food products widened, first, because there were many more mouths to feed, and second, because each mouth demanded a more varied and expensive diet.

Now let us take a more careful look at the shifters of demand that have been at work in the first half of the twentieth century. And let us look first at the demand shifter, change in income. Sometime in our national history, the income elasticity for food fell, and fell drastically—probably during the decades preceding and following 1900. In other words, I posit here that real income increases for average consumers were so great during this period (about 100 percent between 1880 and 1920) that to an important degree the average consumer broke through that income range where rising incomes shoot into purchases of more food and more expensive food and moved into that income range where changes in income have little effect on total food consumption. In any event, the income elasticity for farm food products is now in the neighborhood of 0.2—meaning that the average person's consumption of *farm food products* increases 2 percent with a 10 percent increase in income. Consumers in the 1950s prefer to use additional income to purchase automobiles, durable goods, sporting goods, vacations, and services with their food rather than more food.

In this instance we are *not* talking about the income elasticity of food items purchased in retail stores; the income elasticity of food that consumers purchase at retail runs about 0.6 to 0.7, and it is this high because the income elasticity for nonfarm food services associated with, or built into, those food items (storing, transporting, packaging, processing, and merchandising) is much higher—running between 1.0 and 1.3. In less technical language, consumers are ready and eager to buy more conveniences and gadgets (e.g., TV dinners) along with farm food products as their incomes rise. But they are not so willing—in fact they are reluctant—to buy more farm food products as their incomes rise.

The income elasticity of farm food products is not likely to approach zero in the late 1950s or early 1960s, but it may approach zero by 1975, and if not by 1975 then certainly by the year 2000. The income elasticity for farm food products cannot fall immediately because in 1955 there were some 60 million consumers in the United States, living in families and as single individuals, with incomes of less than $3,500. These are the consumers who currently increase their consumption of animal products and fresh fruits and vegetables importantly as their incomes rise. This is the group that has pulled the income elasticity for farm food products for the average consumer up to as high as 0.2 in the 1950s.

But if per capita real incomes increase as much from 1955 to 1975 as they did from 1935 to 1955, the poorest family in 1975 will, relatively speaking, be living in grand style in that not too distant year. That is, barring a major economic depression, we can look forward to a time, perhaps by 1975 and certainly by 2000, when the basic wants of all families in the United States will have been satisfied with respect to farm food products—when the income elasticity of those products will have fallen to zero. This is not to say that the income elasticity for *nonfarm* food services will have fallen to zero. On the contrary, this latter elasticity may remain above 1.0 as more and more wives escape from the kitchen and experience the joys of dining out.

The key point to keep in mind in all this is that consumers purchase two very different categories of resources in what is commonly called food: farm resources embodied in farm food products, and nonfarm resources converted into services associated with and built into farm food products. The consumption of the second category of resources increases at least proportionately with increases in income, but the consumption of the first category of resources increases only modestly with increases in income, and may in the foreseeable future cease to increase at all.

One qualification needs to be added to this conclusion, and it is not a happy one. Should the economy of the United States run into a prolonged depression, the income elasticity estimate of 0.2 indicates the contractions in the purchases of farm food products that would occur with declining incomes, *until personal incomes began to fall seriously*. But once real personal incomes declined to levels substantially below those of the 1950s, the income elasticity for farm food products might be expected to rise; and a higher income elasticity would, of course, accelerate the contraction in the aggregate demand for food. So it turns out that rising real incomes now do little to expand the demand for farm food products and may in the future do nothing; but falling real incomes under depressed economic conditions could contract the aggregate demand for food, and contract it importantly.

Now we turn to the second demand shifter, population growth. Fortunately for the agricultural sector, developments with respect to population growth as a demand shifter have not paralleled those of rising real incomes. For a while it was feared that population growth also was losing its power as a shifter of demand; in the 1930s most predictions of population growth had the population of the united States leveling off and declining in the 1960s. But for some totally unexplainable reason, during and following World War II the people of the United States decided to wreck the prewar population projections of demographers by going on a child-producing spree. The rate of population increase in the United States in the 1950s is among the highest in the world—15 per thousand as compared with 4.5 for the United Kingdom, 13 for India, and 20 for China.

In comparative terms, perhaps more relevant for this discussion, the total population of the United States increased nearly 9 percent in the decade ending in 1935, 9 percent in the decade ending in 1945, and 16 percent in the decade ending in 1955. And if the 1954–55 rate of increase is maintained, the country's total population will increase by some 16 percent in the decade ending in 1965, to reach 193 million, and by 37 percent in the two decades from 1956 to 1975, to reach 228 million. Even if the 1954–55 rate of population increase moderates slightly, as many experts believe it will, there are still going to be many more people around in 1975 to be fed—certainly no fewer than 210 million.

The picture that emerges with respect to further expansions in the demand for food looks like this: The United States is approaching that state of opulence where further increases in real personal incomes will not expand the demand for food. Expansions in the aggregate demand for food are becoming dependent on population growth alone. But during 1955–75, population growth *is* going to be a powerful force increasing the demand for farm food products.

One more idea should be incorporated into this discussion of the aggregate demand for food. As this aggregate demand relation expands through time, driven by population growth, it will in all probability become even more inelastic than it is now. Further decreases in the price elasticity of the aggregate demand for food will result from further increases in real personal incomes. Rising real incomes historically have had the effect of reducing the price elasticity of the aggregate demand for food, and there is every reason to believe that this tendency will continue until food, like safety pins now, has a zero price elasticity. When this happens—and the time is not too far off—if the fantastic rate of economic development of 1946–56 is maintained, the aggregate demand for food will expand in a perfectly inelastic manner, powered by population growth alone.

SUPPLY SHIFTERS

During the nineteenth century, total farm output in the United States increased as the result of two basic forces: farm technological advance and an increase in the size of the total fixed plant—an increase in the number of acres incorporated into going farms. The former was a minor cause and the latter the major cause. Regardless of the level of farm prices or the pattern of commodity prices, settlers pushed back the frontier year after year and added land and farmsteads to the total fixed plant. Land settlement was part of the great westering movement in America over three centuries, and though it may have grown out of certain economic dislocation and adjustments in the Old World, it was in no way related to the pricing system, except insofar as hard times in the cities forced more people onto the frontier.

With the turn of the century, this means of expanding total farm output began to disappear—the country was almost settled. And by 1920, except for some irrigation developments in the arid West, the total agricultural plant stopped growing. Total farm output did not, however. As already noted, total farm output increased some 90 percent between 1914 and 1956. This great increase resulted almost exclusively from technological advance on existing farms—a process that gained momentum in the latter part of the nineteenth century, particularly in the form of farm mechanization, and became an all-inclusive, ever-present force by the end of World War I. The minor force behind output expansion in the 1800s became the major force in the 1900s.

The development and adoption of many new technologies have contributed to the expansion of the total marketable output of farm commodities since 1920, but no technology has been so important as the tractor. The substitution of tractor power for animal power has released some 70 million acres, or one-fifth of our cropland, for the production of marketable crops. And the myriad hookups that have been developed to go with the tractor have released several million workers for nonfarm employment (a development that is sometimes considered a mixed blessing). Certainly the gasoline engine in its many forms and uses dominated farm technological advance between 1920 and 1950.

Now, there are some who say, because they see evolving no single technology comparable in importance to the tractor, that farm technological advance too is losing steam as a shifter of the aggregate supply relation. But people who take this view don't appreciate how far the development and application of new technologies now permeate the agricultural scene. True, no single developing technology dominates the scene in the 1950s as the tractor did in the previous three decades. But one dominant technology is not a requirement of rapid technological advance and hence of a rapid rate of output expansion. Total farm output is expanding rapidly in the 1950s as the result of technological development and adoption on many fronts: plant and animal breeding; plant and animal disease control; feeds and feeding practices; water control and usage; and soil, crop, and animal handling equipment. Advances on all of these technological fronts contribute to a situation in which the typical farmer adopts one to several new practices each year, thereby becoming more productive.

In this connection we should recognize, too, that the development of new technologies and their adoption on farms is no longer left to chance. The outpouring of new production practices and techniques that occurred between 1920 and 1955 was not the work of a few lonesome inventors in attics or old barns; rather, it resulted from organized and well-financed research at several levels: pure research on natural phenomena, applied research on agricultural problems, and commercial research on specific products and techniques.

Private and public agencies currently spend up to $200 million each year, and perhaps more, on the development of new technologies for *farm production*.[3] And all the efforts directed toward carrying these new technologies to farmers must run into sums much larger than that. Consider all the private and public agencies involved: the federal-state extension service; the vocational agricultural teaching program in high schools; the soil conservation service; the service work of farm cooperatives; and finally and probably now most important, the selling and service work of private feed firms, machinery and equipment firms, processing firms, and producer associations.

Adopting new production techniques has become as much a part of farming as getting up in the morning: farmers are "expected" to adopt new practices and technologies that reduce their costs and expand their output in the same way that they are expected to care for their livestock, send their children to school, and honor their spouses. The farmer is the last link in an endless chain of events, called technological advance, that almost everyone considers good.

Those folks looking for (or trying to hide from) a second revolutionary development to follow the tractor may find it before the turn of the next century. If and when artificial photosynthesis breaks out of the laboratory, where it is now an established fact, that institution-shaking technology will have arrived. When we can produce carbohydrates directly from the sun's rays, without the work of plants, then the production of food can be transferred from farms to factories, and the greatest of all agricultural revolutions will rage across the land. And artificial photosynthesis is certainly further advanced in the 1950s than was the unleashing of the atom in 1900.

So we conclude that the development of new and improved technologies, and their adoption by farmers, gives every evidence of remaining a powerful force in agriculture, driving the aggregate supply relation before it in an expanding action. And truly revolutionary developments are in the offing that could turn present-day agriculture upside down.

The Race, 1955 to 1975

The long-run race between aggregate demand and aggregate supply thus, for all practical purposes, turns out to be a race between population growth and farm technological advance. But since nobody is omniscient, it is impossible to demonstrate that population growth will outrun technological advance between 1955 and 1975, or the converse. An observer who is more impressed with the capacity of Americans to reproduce themselves than with their ability to create new ways of producing goods and services will probably conclude that population growth will win the race. But one who is more impressed with their inventive genius and ability to adopt new technologies will probably bet on technological advance.

Which one wins is terribly important to American farmers. If population growth outraces technological advance, other things being equal, aggregate demand will press against supply and push the level of farm prices upward, as occurred between 1895 and 1915. But if technological advance outraces population growth, aggregate supply will press against demand and drive farm prices downward, as has been the tendency since 1948.

Some evidence can, however, be adduced about the outcome of the race between aggregate demand and aggregate supply over the period 1955–75, where conclusive proof is impossible. During 1951–56, total population in the United States increased by exactly 9 percent. At the same time the total output of marketable farm products increased by 13 percent. The figures in this comparison change somewhat depending on the exact years chosen and the output index used, but the general picture does not change. The total output of farm products in the first half of the 1950s is outracing population growth. And this increase in total farm output occurs in the face of a declining farm price level and with no significant increase in the total inputs employed. In the 1950s output expansion is winning the race.

James T. Bonnen, looking forward to 1965 in a major study that assesses the output-expanding potential of all "known and almost known technology," suggests that the trends of the early 1950s will not be reversed.[4] Assuming that the farm price level is maintained at the 1955 level, which relatively speaking is low, Bonnen estimates that total agricultural production will increase by 30 percent between 1955 and 1965. Taking into account an estimated 15 percent increase in population over the period and a 4 percent increase in per capita food consumption, the Bonnen model predicts that the annual rate of farm surplus, which stood at 8 percent of total supply in 1955, would be enlarged to 12 percent as of 1965. In other words this study, taking a comprehensive look to 1965, concludes that output expansion will increase its lead over demand expansion in the years ahead. In terms of the 1955 farm price level, the total farm surplus will increase from 8 percent in 1955 to 12 percent in 1965.

It is my judgment that the rate of aggregate output expansion can easily exceed the rate of aggregate demand expansion over the period 1955-75. And what can easily occur probably will. In this probable event, one of two things must happen: the annual accumulation of surplus stocks by government must increase or the farm price level must fall. In other words, I believe that the expanding supply relation will press against the expanding demand relation in the long-run race between the two and either drive farm prices lower or, at a supported price level, drive more stocks into the hands of government. The capacity to expand farm output beyond the needs of the population is there, and unless counteracted in some effective way, this capacity will further intensify the general income problem in agriculture.

All that I have said to this point with regard to the race between the aggregate demand for food and the aggregate supply over 1955–75 has assumed an expanding total economy with full, or nearly full, employment. In other words, implicit in the discussion up to now has been the assumption of continued prosperity. This appears reasonable, but it is only an assumption. A slowing down of total economic growth is a distinct possibility, and a major and prolonged economic depression is at least possible during the period.

The principal short-run effects on agriculture of a business depression are a slowing down—and possibly even a contraction—of the rate of aggregate demand expansion, by reason of a decline in real personal incomes, and a drying up of job opportunities for surplus agriculture workers in the nonfarm sector. Both of these effects work in the direction of pushing down farm prices. The longer-run effects on agriculture of a general economic depression might be a slowing down of the rate of population growth and a slowing down of the rate of farm technological advance. These effects would tend to cancel each other out, but to what extent it is impossible to say.

One conclusion can be reached. An economic depression before 1975 would have the immediate effect of slowing down the rate of aggregate demand expansion *relative to* the rate of aggregate output expansion and hence act to drive farm prices to an even lower level than that indicated under continued prosperity. The farm price-income problem in the latter half of this century will be more than difficult to solve under conditions of prosperity; under depression conditions it would be impossible.

Market Organization and Technological Advance

Why in the face of falling farm prices and declining gross incomes do farmers persist in adopting new technologies, and thus expanding output? Why, in the 1950s, have farmers pushed aggregate output ahead of demand through widespread technological advance and thus driven down the prices of their own products? And why are they likely to keep right on behaving in this seemingly irrational manner? In the main, the answer is to be found in the market organization of agriculture. But given this market organization, we need to consider some other factors: the role of society acting through government, and the financial position of farmers. So let us inquire into the way farmers adopt new technologies to see how and where they are led astray.

MARKET ORGANIZATION AND THE ADOPTION PROCESS

To this point farm technological advance has been considered in terms of the total agricultural industry—in terms of its shifting effects on the aggregate supply relation, hence on the farm price level.[5] In this view, we see the effects on the industry of a particular technology or production practice *after it has been widely adopted throughout the industry*. If, however, we take as a unit of analysis not the industry but rather different firms in the dynamic process of adoption, the story changes. And this is what we shall do now: consider the effects of farm technological advance on firms that adopt the technique early, then on the more typical followers, and finally on the laggards.

We should recognize, first, that farmers typically operate in a special sort of a market, one that satisfies the key conditions of a perfectly competitive market—a market in which no one farmer can have, or does have, any perceptible influence on the price of his product (or his factors of production). The farmer is a price taker; he takes the price offered him because he is such a small part of the total market that he can have no perceptible influence on the market or on the market price.

Second, recall that a technological advance has the effect of lowering the *per unit costs of production* of the farm firm (typically, per unit costs of production are reduced as the total value of the product increases by more than the increase in total costs; it is extremely difficult to think of a new technology that does not increase output). This being the case, farm producers who adopt a new technology (e.g., growing hybrid seed corn) early in the game realize increased net returns from undertaking that enterprising act. The new technique reduces costs of production for the enterprising few, but they are such a small part of the market that total output is not increased noticeably and price does not come down. Net incomes of the few adopters increase, creating a powerful incentive for other farmers to adopt the technique.

In this explanation we find the basis of continuing and widespread farm technological advance. The operators who first adopt a new technology reap the income benefits (the difference between the old price and the new, lower unit costs). Then other farmers in the community see the income advantage accruing to Mr. Early Bird; also, the

Extension Service and other educational units spread the information around. So Mr. Average Farmer decides he will adopt this cost-reducing technique, and this includes most farmers in the community. *But the widespread adoption of this new technology changes the entire situation. Total output is now increased, and this increase in the supply of the commodity lowers its price.* And where the price elasticity of demand at the farm level is less than −1.0 (i.e., demand is inelastic), as is commonly the case in agriculture, gross returns to the producers must fall. Further, over time, any increases in net returns are capitalized back into the value of the land, with the result that land prices rise.

As the dynamic process of technological adoption unfolds, we see two things happening: in those cases where the output of the commodity is increased, the price of the commodity falls relative to that of its substitutes; and unit costs of production rise after their initial decline as the gains from the new practice or technique are capitalized into the value of the fixed asset involved. So in the long run, by the time most farmers have adopted the technology, the income benefits that the first farmers realized have vanished. Mr. Average Farmer is right back where he started, as far as his income position is concerned. Once again, average unit costs of production are equal to price, and no economic surplus remains.

If this is the typical result, why do farmers generally adopt new methods? It is easy to see why the first farmers undertake a new method or practice. They benefit directly. And we can understand why neighbors of the enterprising first farmers adopt the technology: they see the income advantage and make up their minds to give it a try. But as more and more farmers adopt the new technology, output is affected and the price of the commodity declines. This price decline acts as a burr under the saddle of the followers, the average farmers; the price of their product is declining, but their unit costs of production are unchanged. *To stay even with the world, these average farmers are forced to adopt the new technology*. The average farmer is on a treadmill with respect to technological advance.

In the quest for increased returns, or the minimization of losses, which the average farmer hopes to achieve through the adoption of some new technology, he runs faster and faster on the treadmill. But by running faster he does not reach the goal of increased returns; the treadmill simply turns over faster. And as the treadmill speeds up, it grinds out more and more farm products for consumers.

In a sympathetic book in which he tells the story of the tides of agrarian revolt in the United States, with a rough-and-ready sense of humor Dale Kramer calls those farmers who revolt the wild jackasses.[6] Perhaps it is appropriate in this context to say that the many farmers running on the agricultural treadmill have been used as tame jackasses—grinding out greater and greater supplies of food at no advantage to themselves, and perhaps to their disadvantage.

The position of the laggard who will not or cannot adopt the new technologies is a tragic one. The farmer who belongs to a religious sect that does not permit technological advance, the aged or beginning farmer who cannot afford the initial cost of the technology or production practice, or the lazy fellow who prefers to go fishing, finds himself in an income squeeze. The relative price of the commodity falls as one technique after another is adopted throughout the industry, but his unit costs of production do not come down. Thus, the farmer who does not adopt new technologies and practices is squeezed and squeezed. Farm technological advance for him is a nightmare.

In the preceding analysis, we reached the conclusion that any economic surplus growing out of the introduction of a new technology is squeezed to the zero point in the long run. But this does not mean that the labor income to operators and hired labor must fall, even when the price level is falling. We have yet to take up another dynamic consideration: the substitution of machinery and equipment for labor. Increased farm mechanization most often takes the form of substituting a machine process for a hand process. If in these situations gross returns are unaffected and total costs of the unit under consideration (the firm, or total agriculture) decline, then the labor income of those workers remaining *must* be increased.

In the more typical case in which output expands, demand is inelastic and gross revenue declines. If the impact of the new technology (e.g., the general purpose tractor and the variety of hookups) reduces the number of workers required in the industry, as well as reducing total costs of production, the average labor income of those remaining *may not* decrease; or if it does it will not decrease as much as it otherwise would. In this situation, the average labor income of those remaining in agriculture develops into a three-cornered race between declining total revenue, declining total costs, and declining numbers of workers. No generalization can be made with respect to this race except to say that it probably operates to increase income disparities *within* agriculture: the labor income of the efficient producer, farming more land with more capital, holds constant or rises, and the labor income of average and poor farmers declines.

SOCIAL ACTION AND THE ADOPTION PROCESS

The typical small family farmer, of course, is not and has not been in a position to undertake the costly, time-consuming work of developing new technological practices for the farm operation. But the many small farmers who make up the agricultural industry have rarely organized to promote and finance research and development through their own private agencies. In fact—and even though farmers are notorious tinkerers—farm people have often displayed antagonism toward scientific inquiry and development. In recent years they have come to accept and even rely on a continuing change in the state of the art in agriculture, but in most instances they have not initiated those changes. Thus we conclude that if the availability of new production practices and techniques were dependent on farmers' initiative, they would have been in short supply for many years.

But new production practices and techniques for agriculture have not been in short supply; on the contrary, there has been a continuous outpouring of these new technologies in the twentieth century. And the most important reason there has been this generous supply since the turn of the century is that society decided to take collective action to ensure this ample supply of new technologies.

The nation established "agricultural and mechanical arts" colleges in the mid-nineteenth century to service the technological needs of agriculture. Society, acting through the federal government and the various state governments, has more than generously financed the research and development work in those colleges and in government research agencies ever since. This is not to say that every important technological development in agriculture has been the direct result of work in the land-grant colleges and government research agencies. Far from it: private agencies have contributed many new technologies for use in agriculture, and they appear to be providing an increasing proportion. But it is to say that *society has covered the overhead costs of training scientists and carrying on the basic research that lies behind every applied technique.*

The point is that society has underwritten technological advance in agricultural by guaranteeing a continuous outpouring of new production techniques for adoption on farms. If farm technological advance does not outrace population growth in 1955–75, it won't be because the new techniques are not there to be adopted. By generously financing research and development in agricultural production, society has made it as certain as possible that an ample supply of new techniques will continue.

This willingness of society to finance research and development in agricultural production is in many ways a strange affair. Perhaps in some collective and intuitive sense society feels that rapid technological advance in agriculture is basic to rising levels of living for its members—as indeed it is. By underwriting a rapid rate of technological advance, society assures itself of a bountiful food supply at relatively low prices. But the strange aspect of all this is that this generous financing of research and development is done in the name of helping farmers, and it is so accepted by most farmers and their leaders.

Now, in the short-run monopoly sense (i.e., the highwayman sense), nothing could be further from the truth. The monopolist always seeks a position where his produce is relatively scarce and the products of all other groups are plentiful; from this position of market power the monopolist trades scarce, expensive items for plentiful, cheap items. But a rapid rate of technological advance—a rate of technological advance that drives aggregate supply ahead of aggregate demand—places farmers in just the opposite position. It puts them in the weak market position of producing bountiful supplies at low prices.

Farmer Asset Positions and the Adoption Process

In a free market, at least, aggregate supply cannot outrace aggregate demand indefinitely. At some point its pace must slow down and become equal to, or perhaps even lag behind, the rate of demand expansion. In other words, the expansion rates of these two relations are related; the connection is somewhat indirect, but it is there. The aggregate demand and the aggregate supply relations are related through the nexus of the asset positions of farmers.

Most new technologies adopted on farms are capital using—that is, their adoption requires an additional cash outlay or some kind of additional financial commitment. But since adopting a new technology reduces unit costs, farmers are willing to make the additional investments so long as they can. And they can so long as their liquid and capital asset positions are strong and unimpaired.

But these asset positions deteriorate under a falling farm price level with the attendant declines in gross and net incomes. The liquid asset position typically goes first; but with the passage of time capital assets become encumbered in order to cope with losses resulting from declining incomes. In a free market situation, then, farm technological advance sows the seeds of its own slowdown. If aggregate output outpaces aggregate demand long enough and far enough, and if the farm price level falls far enough and stays down long enough, the asset position of farmers generally will become weak, and the adoption of new technologies must be choked off. In this way the rate of output expansion is slowed down and brought into equality with the rate of demand expansion.

This, of course, is a painful process, as farmers have discovered in a limited way in the 1950s. Furthermore, it is no simple process. The slowdown in farm technological advance will not be uniform. It will first strike the inefficient farmers and the beginners: their vulnerable asset positions deteriorate rapidly when the farm price level declines. The average farmer will hold out longer, and the very efficient farmer succumbs to the slowdown only in extreme situations. The efficient farmers, the early adopters of new techniques, who reap the income rewards accruing to them as such, and who successfully increase their labor income by substituting machinery and equipment for labor, can withstand and even thrive in a major price decline *to a point*.

I should emphasize here, however, that the rate of output expansion, powered by farm technological advance, does not slow down immediately when it encounters a price level decline. Witness the rapid output expansion in the early 1950s in the face of falling farm prices. Farmers generally came out of the World War II period with strong asset positions and for this reason generally have been able to maintain a rapid rate of technological advance in spite of falling prices. The rate of output expansion in the 1950s and 1960s will not slow down until the asset positions of these many average or representative farmers begin to be impaired. But in a free market situation the slowdown in the rate of aggregate output expansion must ultimately come, as the slowdown in the adoption of new technologies engulfs more and more farmers.

The question may be asked, Will the rate of aggregate output expansion be slowed down and become equal to the rate of demand expansion by 1975? The answer must be: It depends on the kinds of government action taken—on the collective action of society. If farm prices and incomes are supported in some way but production controls remain ineffective as in the past, then the answer is probably no. At the 1955 farm price level, with production controls no more effective than those in existence before 1955, the evidence suggests that aggregate output will continue to outdistance aggregate demand. But a return to a free market in agriculture and the lower level of farm prices that would be generated in a free market would probably lead to a decline in the rate of output expansion and possible equality with the rate of demand expansion by 1975, through the nexus of farmers' asset positions.[7]

The way of agriculture may be hard, but it is not hopeless, as the conclusions above might imply. Those two policy alternatives do illustrate, however, the effect that farm asset positions generally may be expected to have on the rate of aggregate output expansion.

The Consequences of Farm Technological Advance

The spotlight in this section will be on 1920–55, the period when total inputs employed in agriculture remained constant and when increases in total output must therefore have resulted from new configurations in the use of productive resources (from technological advance). Recall that total farm output increased throughout the nineteenth century and during the first two decades of the twentieth as the result of an expansion in the size of the fixed plant of agriculture and technological advance, with the former decreasing in importance and the latter increasing over the long period. Hence it is difficult, if not impossible, to know what part of the increase in total output is attributable to technological advance and what part to an increase in the size of the fixed plant during that long period. But from 1920 to 1955 all, or practically all, of the increase in total marketable output must be attributed to technological advance. There is no other explanation.[8]

FOOD SUPPLY

The technological developments that occurred during 1920–55 and were adopted on commercial farms enabled farmers to produce with the same total volume of resources a food supply more than adequate for the growing population. Thus the first and most necessary goal of this and every society was achieved: an adequate food supply. Never before had an adequate food supply for an expanding population been provided without employing more total resources in agriculture. John M. Brewster sums up the consequences of farm technological advance for the food supply in these words: "For the first time in history, gain in labor productivity appreciably outraced the rate of population increase during the 20's and became four times faster during the 30's. For every 100 workers needed in agriculture in 1930, only 79 were needed in 1940. The pressure of population upon the food supply had done an about face."[9]

First through the beneficence of plentiful and fertile land and second because of technological advances, consumers in the United States enjoy a rich, varied, and, if they so choose, nutritious diet. This is a blessing that few people in the past have enjoyed and that relatively few people in the world enjoy even today.

RELEASE OF HUMAN RESOURCES

Not only does farm technological advance assure Americans of an adequate food supply in the 1950s, but over the years it has released millions of farm-reared people to work in manufacturing, in the distribution system, and in the arts, sciences, and professions. This, of course, is the mark of economic progress: first the release of workers from agriculture to go into manufacturing, and then their release from both of these categories to enter the service trades (and to enjoy increased leisure time) *as the real incomes of all continue to rise.*

The process by which these people have been released from agriculture has not always been kindly, but it has taken place. It has happened as workers in agriculture have become increasingly productive—as one worker, armed with new production techniques, has been able to produce enough food and fiber to meet the wants of more and more nonfarm people. To illustrate: in 1820 one worker in agriculture could support 4.12 persons including himself or herself, and by 1920 one worker could support 8.27 persons; but by 1955 one worker in agriculture could support 19.74 persons. In other words, the capacity of an agricultural worker to feed and clothe people doubled in the first hundred years but more than doubled in the next thirty-five.

The proportion of the total labor force of the United States employed in agriculture has declined steadily over the long-run past, from 72 percent in 1820 to 18 percent in 1920 to about 10 percent in 1955. This is common knowledge. But not so generally known is that total employment in agriculture reached a peak of 13.6 million workers in 1910 and has been declining ever since. Further, the movement of labor resources out of agriculture has been rapid in recent years: total employment in agriculture fell from 11 million persons in 1940 to 9.3 million in 1950 and to 8.2 million in 1955. This is no small decline.

Since 1920 the substitution of machines for men in agriculture has been rapid indeed. Where this substitution will end no one knows, but it is reworking the face of agriculture in the 1950s, increasing the average size of farms (in acres), greatly increasing the capital investment on farms, and commercializing the farm operation. And the substitution of machines for humans may reorganize the family farm out of existence in years to come.

Farmers' Incentive Incomes

Incentive income is a concept developed by J. R. Bellerby to describe the return to human effort and enterprise.[10] In farming this is the return to the farmer as a manager, laborer, and technician. It does not include any return to property or capital. The income incentive ratio relates the incentive income of farmers on a man-unit basis to the incentive income of persons engaged in nonfarm enterprises on a man-unit basis. The incentive income ratio thus compares the average, or per unit, return to human effort and enterprise on the two sides of the farm/nonfarm fence. By five-year intervals, the farm/nonfarm incentive income ratio for the interwar period is as follows:[11]

1920	46
1925	38
1930	32
1935	43
1940	32

During the interwar period the incentive income of farmers is consistently less than 50 percent of nonfarm incentive incomes. In other words, when only the returns to human effort and enterprise are considered (i.e., when the returns to capital are excluded), the returns to farmers as compared with nonfarm workers are shockingly low.

It is interesting that Bellerby computed these incentive income ratios for many countries for the interwar period, and that the United States falls near the bottom of the list. The incentive income ratio is higher for all the following countries than for the United States: Australia, New Zealand, France, the United Kingdom, Denmark, Sweden, and Canada. The incentive income ratio of the United States exceeds that of only a few developing countries—Egypt, Mexico, and Thailand, for example.[12]

As we have observed, widespread technological advance on farms in the United States since 1920 has assured the growing population of the United States of an adequate food supply and gives promise of underwriting an adequate food supply in the future. Widespread farm technological advance has also released several million agricultural workers for nonfarm employment and gives promise of releasing many more. But widespread farm technological advance during the interwar period did not result in high incentive incomes for farmers relative to nonfarm workers. It contributed to low incentive incomes for farmers as farm technological advance first expanded aggregate output and then sustained aggregate output in the face of a contraction in aggregate demand during the Great Depression. In the

United States, then, where farm technological advance was the most rapid in the world during the interwar period, the incentive income ratio was among the lowest in the world.

Immediately following World War II, when the United States made some effort to feed the hungry, famine-ridden peoples of Europe and the Far East, the income incentive ratio rose to the 50 percent level. But with falling gross and net farm incomes in the 1950s resulting from the falling farm prices level that resulted, in turn, from aggregate output marching ahead of aggregate demand again, the incentive income ratio fell back to the levels of the depressed 1930s. The incentive income ratio is given below for selected years:[13]

1947	50
1950	44
1955	30

Truly, American farmers are on a treadmill. They are running faster and faster in the quest for higher incomes growing out of the adoption of new and more productive techniques, but they are not gaining incomewise. *They are losing.*

The General Theory of the Agricultural Treadmill

The capacity of American farmers to command good and stable prices and incomes in the market is weak; their power position in the market is weak. The farmer *takes* the prices the market offers him or her, and very often these are low prices.

The farmer's weak position in the market (i.e., the low prices offered him, and his inability to reject those low prices and command higher ones) grows out of three related circumstances: first, the high value that American society generally places on technological development and application; second, the market organization within which farmers operate; and third, the extreme inelasticity of the aggregate demand for food. The combination of these circumstances places farmers in an unenviable position.

The American people have not singled out agriculture to carry the burden of technological advance; Americans prize technological advance, expect it, and demand it in all segments of the economy. As Bushrod W. Allin stated in the 1957 Lecture Series of the Graduate School of the United States Department of Agriculture, *belief in technology* is a part of the American creed: "[The] dynamics in American culture I shall call the American creed—a blend that is peculiarly American. The three principal dynamics of this creed are: 1. Belief in enterprise; 2. Belief in democracy; 3. Belief in technology."[14] Americans are willing to back this belief in technology with dollars—7 billion of them in 1956.[15] With few exceptions, businessmen believe it is good business to develop new and better products, and to this end they spend vast sums in research and development. In fact, competition in the nonfarm sector commonly takes the form of product competition; in this common situation firms do not compete through price; they compete through product differentiation— by means of an improved product or a different product. As we have already noted, society has been generous in financing research and development in agriculture. Our society expects a rapid rate of technological development, and it has experienced a rapid rate of technological advance in most lines of endeavor, including agriculture.

The farmer operates in a sea of competitive behavior; each farmer is a tiny speck on this sea, and the output of each farmer is a tiny drop in this sea. With rare exceptions, the single farmer operates in a market so large, that he can have no perceptible influence on it. In this situation, the farmer must take as given the prices generated in the market.

Confronted with this situation, he or she reasons, "I can't influence price, but I can influence my own costs. I can get my costs down." So the typical farmer is always searching for some way to get his costs down. By definition a new technology is cost reducing (i.e., it increases output per unit of input). Thus the farmer is always on the lookout for new, cost-reducing technologies. Built into the market organization of agriculture, then, is a powerful incentive for adopting new technologies—the incentive of reducing costs on the individual farm.

Now, if the demand for food were highly elastic, all would be sweetness and light in agriculture. If the aggregate demand for food were *elastic*, the bountiful and expanding supplies of food that farmers want to produce would sell in the market at only slightly reduced prices, and gross incomes to farmers, in the aggregate and individually, would increase. But the aggregate demand for food is not elastic; it is inelastic, and extremely so. For this reason, a little too much in the way of total output drives down farm prices dramatically and reduces the gross incomes of farmers in a similar fashion. Furthermore, the persistent pressure on each farmer to adopt new technologies and thereby reduce unit costs has the effect of constantly putting a little too much in the way of supplies on the market. The peacetime tendency for aggregate supply to outpace aggregate demand keeps farm prices relatively low.

I have sketched a general theory of the agricultural treadmill. The high value that society places on technological advance guarantees a continuous outpouring of new technologies. The incentive to reduce costs on the many, many small farms across the country guarantees a rapid and widespread adoption of the new technologies. Rapid and widespread farm technological advance drives the aggregate supply relation ahead of the expanding aggregate demand relation in peacetime; and given the highly inelastic demand for food, farm prices fall to low levels and stay there for long periods.

3 | FARM TECHNOLOGY, FOREIGN SURPLUS DISPOSAL, AND DOMESTIC SUPPLY CONTROL

This essay was my presidential address to the American Farm Economics Association in 1959.[1] I argue that we have the opportunity to convert American agricultural abundance in the form of burdensome product surpluses into productive resources to aid economically backward countries in their development efforts. To turn this opportunity into a reality, the United States must move away from short-run foreign disposal schemes to sustained, innovative development projects for the recipient countries. And for American farmers to benefit incomewise from this policy approach, or for that matter from any other that expands the demand for their products, they must accept effective production control measures.

President John F. Kennedy and Secretary of Agriculture Orville Freeman were favorably disposed toward this policy approach, but it was rejected out of hand by the powerful conservative forces in Congress.

The Technological Revolution

During the 1950s total farm output has increased approximately 2.5 percent per year as total population in the United States has increased some 1.8 percent per year. Population growth speeded up in the 1950s, but so did the rate of output expansion. At this time, the aggregate output of agriculture is outdistancing a very rapid rate of population growth in the United States by more than 0.5 percent per year. And where the income elasticity of raw farm products approaches zero, as it does in the United States, this imbalance can be disastrous. Where the income elasticity of raw products approaches zero, the excess rate of output expansion over population growth properly measures the additional pressure of supply on demand each year; hence it measures the increased downward pressure on farm prices (or, under price supports, the widening of the annual rate of surplus).

And the surpluses have mounted. Of this development the secretary of agriculture has kept us well informed. The U.S. Department of Agriculture's investment in storable stocks has increased persistently—from under $2 billion in fiscal 1952 to an estimated $9 billion in fiscal 1959. And this buildup of stocks has occurred in the face of really massive export programs—programs that have moved between $1.5 and $2 billion worth of agricultural commodities, in addition to conventional commercial sales, in each of the past four years.[2] Think where Mr. Benson's stockpile would have been (or where farm prices could have been) without this volume of foreign surplus disposal. Further, this mounting stockpile is not limited to wheat and cotton; it is now heavily weighted with feed grains, which we all recognize as unprocessed livestock and livestock products. In sum and in short, the surplus condition is general; the mounting surplus in feed concentrates makes it so.

Now, what has pushed the rate of output expansion ahead of population growth by 0.6 to 0.7 percent per year? Basically it is the rapid and widespread application of new knowledge to agriculture—new knowledge that is expanding output faster than offsetting adjustments can be made in the way of reducing the employment of other factor inputs in agriculture, notably labor. The rain of new knowledge across the land, the technological revolution sweeping over agriculture, is not a narrow thing tied to machinery and equipment—it is a broad thing involving improved skills in labor and management, the relocation, recombination, and area specialization of commodity enterprises, and the farm adoption of new techniques. All these avenues of new knowledge application, acting and interacting, are raising production functions, lowering cost functions, and expanding output in agriculture.

Preliminary estimates from the U.S. Department of Agriculture suggest that total inputs in agriculture increased about 10 percent between 1940 and 1958 while total output increased 50 percent. And since I have seen no evidence of increasing returns to scale in agriculture, after the obvious smaller than one man unit is passed (in fact, all the evidence I have seen suggests constant returns to scale), the inter-pretation of these data must be that 20 percent of the increase in total output since 1940 is explained by an increase in inputs and that 80 percent of the increase is to be explained by technological advance in the broad sense outlined above.

There is, however, a new school of thought emerging—or perhaps an old school gaining a new lease on life—with respect to the employ-ment of resources in agriculture. Briefly, it says that up to 60 percent of the increase in output since 1940 is to be explained by an increase in total inputs in agriculture. The disagreement seems to revolve around the different views held with respect to proper rates of de-preciation for farm machinery and equipment. Hence we may antici-pate a battle of input indexes in the years to come.

Although this second view of total input behavior in agriculture dampens somewhat the technology-advance thesis of certain earlier writings of mine, it in no way subtracts from the technological-revolution thesis of this paper. In both views the aggregate input of labor in agriculture has declined drastically since 1940—by more than 35 percent. Inputs of land have held fairly constant, and inputs of new kinds of nonfarm capital have increased dramatically. In both views, new and improved capital items have substituted for labor, but in the second view a considerably larger quantity of capital was required to obtain the output increase that occurred than in the first, or Co-chrane, view. The only question at issue is, How much additional capital was required?

Thus, this is the ideal construct I wish to leave with you. New knowledge flows into agriculture in many ways: in the form of de-veloped machines and techniques to be adopted, in the form of new enterprise combinations (industry relocation and specialization), and in the form of increased labor and management skills. Now, the first two of these will be adopted, or instituted, as rapidly as they become available and as rapidly as labor and management skills per-mit. They will be adopted in the pure substitutional case of capital for labor to realize the increased labor-management returns to those persons remaining in agriculture. They will be adopted in the pure technological-advance case first to realize the enhanced profits of the early adopters and second to reduce unit costs on representative farms as product prices fall with expanding supplies. And they will be adopted, or instituted, in practice for a combination of these two reasons. The incentive to apply new knowledge is there: it is powerful and ubiquitous in American agriculture.

At this point one may ask, Will the new knowledge and the new technologies continue to pour forth at a rate that pushes output ahead of demand? This question is most often asked by those taking a narrow view of the technological revolution in agriculture—those who tend to view advancing technology in terms of machines and equipment. But I would argue that this is a too limited—in fact an erroneous—view of the technological revolution in American agriculture. A more correct view, I believe, is that of a broad front of new knowledge flowing into agriculture. Sometimes this new knowledge takes the form of new machines and equipment, but it also takes many other forms. It takes the form of improved disease control, improved pest control, improved water control, and improved breeds and varieties. It takes the form of upgraded labor skills—labor that can handle the newer practices and that can see the place for still different and improved practices. It takes the form of improved management skills that lead to the relocation, recombination, and further specialization of enterprises. And this advancing front of new knowledge is now fed by many streams besides the land grant colleges; it is fed to an increasing degree by private industrial research, and it is becoming an important beneficiary of the large-scale post–World War II research efforts in physics, genetics, and biochemistry. This is serendipity on a grand scale. Thus I can find no reason to anticipate a slowdown in the flow of new knowledge into agriculture. On the contrary, I believe that research and development in agricultural production is running into external economies of scale—is running into and is being fed by the fruits of research in the more basic disciplines (e.g., physics and biochemistry).

In sum, and as I see it, usable knowledge and improved production practices and processes are going to flow into agriculture in increasing abundance. I suspect we are on the threshold of the technological revolution in agriculture, not in the middle or later stages of it. If given half a chance, I can become a real Buck Rogers with respect to the future of agriculture: the feed supply produced in factories employing artificial photosynthesis processes, the sea intensively farmed to yield protein, and animal products produced under controlled conditions such as we are now beginning to see in poultry.

And as we have observed, the incentive to substitute capital for labor and to adopt cost-reducing practices on farms is there. Unless we destroy asset positions of farmers generally by putting them through the long-run wringer of a free market, they are going to continue to substitute capital for labor and to expand output. The twin pressures deriving from the technological revolution in agriculture—the pressure to move labor out of agriculture and the pressure of the food and fiber supply on population—are going to remain with us and intensify because the technological revolution in agriculture is going to stay with us and intensify.[3] The great policy problem of American agriculture in the 1950s and in the decades to come is finding a way to moderate those pressures and make them tolerable to the people living under them—farmers.

The Foreign Disposal Approach

Agricultural exports moving under some kind of special government programs amounted to between 60 and 70 percent of total agricultural exports from the United States in the post-World War II years and continued at these high levels through the hot Korean action. But in the early 1950s—in the first years of the cold war—agricultural exports under special programs fell off sharply; such exports fell from a value of $1.2 billion in 1950–51 to $0.5 billion in 1952–53. This new low level of special-program exports held, however, for only one year. In 1953–54 agricultural exports underwritten by government began moving up, reaching a peak value of $1.9 billion in 1956–57 but leveling off on a plateau of about $1.5 billion in the late 1950s.

The low level of special-program agricultural exports realized in 1952–53 could not hold for the reason discussed above—the pressure of food and fiber supplies on domestic population. Mounting surplus stocks following 1953 forced politicians and administrators to find a way of disposing of those stocks. And the way was found once again in the acceptance of the world's needs for food and fiber as our needs (or at least the needs of the non-Communist part of the world). The pressure of food and fiber supplies on population in the United States was moderated during the late 1950s by massive surplus disposal abroad. To an important degree, we exported our farm problem.

The cynic's view of the motives that led the United States to assume the "burden" of meeting the unsatisfied food and fiber needs of much of the world in the 1950s may not be wide of the mark, but recent efforts by some individuals and agencies to analyze away that world need strike me as both fallacious and malicious. The need is there. Trained observers traveling through the Middle East and much of Asia cannot miss it. They do not have to await clinical examinations of the population to observe it; they can see chronic undernourishment with their own eyes. And the Food and Agriculture Organization has documented this need many times. Using estimates of per capita food consumption and requirements presented in the *Second World Food Survey* and world population estimates for 1956,[4] I get a very rough estimate of the caloric gap of the non-Communist world, *measured in metric tons of wheat*, of some 30 to 35 million metric tons. (This, of course, is not to suggest that the caloric gap *should* be met with wheat alone; *metric tons of wheat* is simply used as a meaningful unit of measure here.) This gap compares with annual wheat production in the United States of between 25 and 30 million metric tons, with the United States stocks of wheat in 1958 of 24 million metric tons, and with the foreign surplus disposal of food and feed grains in 1957–58 of 10 or 11 million metric tons. The need is there and it is large, but it is not out of this world.

Further, it is now pretty clear that the caloric gap in the developing parts of the world is widening rather than narrowing.[5] Per capita agricultural production in the developing countries in the 1950s has not regained pre-World War II levels. Both the Far East and the Near East, which were formerly net exporters of agricultural products, are now net importers. Their changed position is, of course, to be explained by their population upsurge. And the end of this mounting population pressure on food and fiber supplies in these countries is not in sight. Let me sum up and drive home the import of this deteriorating situation with respect to per capita food and fiber supplies in the developing countries by quoting from the recent report of the Ford Foundation agricultural production team sent to India to study the food crisis developing there. The report reads as follows:

> India is facing a crisis in food production . . .
>
> Five million persons per year were added during the First Five Year Plan, and seven million per year will have been added during the Second Plan Period. Ten million per year probably will be added during the period of the Third Plan ending 1966. . . . This explosive increase in population will raise the total from 360 million in 1951 to an estimated 480 million by 1966.
>
> Preliminary planning is now underway for the Third Plan. No specific targets have been announced, but discussions indicate that from 100 to 110 million tons of food grains will be required by 1965–66. . . .
>
> In order to produce 110 million tons of food grains annually by the end of the Third Plan, the rate of production increase must average 8.2 percent per year for the next 7 years. This rate of increase compares with an annual average of 2.3 percent from 1949–53 to 1958–59. The task is overwhelming.[6]

Although the food and fiber needs of the developing countries are great and becoming greater, and although the United States has acted to meet the most acute of these needs, it does not follow that the foreign surplus disposal programs of the United States have in any fundamental sense been good for the recipient countries, for foreign competitors, or for the United States. Many share this negative view; I share it in part myself. Our first efforts at surplus disposal in the 1950s were very crude. We turned our agricultural attachés into order takers; we sent huckster teams around the world to find new markets; we engaged in barter; we pushed our surpluses hard. How much these concessional sales cut into the export markets of such friendly nations as Canada, New Zealand, and Denmark we will probably never know. Probably not as much as was claimed. On the other hand, we were not careful of third-country positions in the mid-1950s.

Toward a Positive Foreign Disposal Approach

But we have learned much with respect to the mechanics of foreign surplus disposal; we have become sophisticated dumpers. We now consult formally with the various export countries through the FAO's consultative committee on foreign surplus disposal; we consult informally with competing nations in the initiation and modification of disposal agreements with recipient countries; and we are firmly committed, in the administration at least, to the "additional principle" (concessional sales, or grants, to a recipient country must represent *additions* to the regular commercial sales of that country).[7] In short, we have made great strides in improving the operation of surplus disposal programs vis-à-vis other nations; also, the sale of food and fiber supplies for foreign currencies is a bright institutional innovation.

But there remain grave difficulties with our total foreign surplus disposal operation—difficulties that in my opinion must be resolved if this course of action is to be pursued on a sustained basis with beneficial consequences for all parties concerned. These difficulties are highly interrelated, but gain in clarity, I believe, through separation. They are:

1. The temporary, emergency quality of our foreign surplus disposal operations. Although everyone likes a bargain, there have as yet been few lasting benefits to the recipient countries. Current consumption levels have been increased, and this can be important to hungry people. But little else has been achieved. Economic development projects, for which currency received from the sale of food and fiber in the recipient countries has been set aside, remain to an important degree in the planning stage; disbursements of funds to finance projects lag far behind the planned use of funds. In short, our foreign disposal programs were conceived by this administration as expedient, temporary disposal measures, and they are so treated by all concerned.[8]

2. The great uncertainty created all around by our foreign surplus disposal operations. We ourselves don't know how long we will place primary emphasis on this type of adjustment, what forms the programs will take, the extent of the price concessions, and the nature of the side conditions attached to such sales. And foreign competitors are completely in the dark with respect to our plans, as are prospective recipient nations. In this context, rational action is impossible. Plans of economic development of several years' duration based on agricultural exports from the United States must be based upon either hope or conjecture; consequently they often are not made. This is the irresponsible aspect of our foreign surplus disposal policy.

The question may then be asked, What do we need to do to make these disposal programs acceptable to friendly competing nations and to contribute to economic development in the recipient nations? I would state the basic need as follows: In disposing of surplus agricultural commodities in foreign countries, the United States must be prepared to make some policy commitments of long-run duration with respect to program objectives, availability of supplies, means of financing, and so on. The specifics of this general course of action can be stated under the following seven points:

1. Except in famine situations, surplus agricultural commodities in the United States, when disposed of abroad, are to be used *exclusively* to finance economic development. Economic development is here interpreted broadly to include education and training of human agents as well as physical capital formation, but all food and fiber shipments must be related to or integrated into a plan, or plans, of economic development.[9]

2. Once "surplus" agricultural commodities from the United States become committed to a development plan or project of a foreign country, for whatever duration, one year or ten, they cease to be surplus commodities, *and the whole operation ceases to be a foreign surplus disposal operation.* At this point the committed supplies become "development supplies" and get built into the aggregate demand for the farm products of the United States; they become a recognized claimant on domestic production in the same sense as the school lunch program and the international wheat agreement.

3. Food and fiber supplies committed to development plans and projects would be financed by means acceptable to the recipient countries, but with the basic objective of speeding economic development: perhaps by grants, perhaps by loans, perhaps by sales for national currencies. The financing principle to be followed here is one that maximizes economic development, not the money return to the United States. The charge that the United States is concerned only with exporting its farm problem would thus be refuted, and the entire program would become a humanitarian program with lasting possibilities.

4. Recipient countries must in every case provide evidence that these "development supplies" from the United States do not reduce their "normal" acquisitions of food and fiber from other countries. By this and the previous point, criticisms by foreign competitors would lose their force.

5. Since "development food and fiber supplies" could finance only a part of every plan or project, complementary programs to finance the purchase of hard goods, construction materials, and services would be necessary. In some cases food and fiber supplies might finance up to 70 or 80 percent of a project (e.g., road building under an antiquated state of the art), but in most cases the percentage would be lower. Hence the financing of nonagricultural supplies must be a part of the total program.

6. Competing nations burdened with agricultural surpluses (e.g., Canada, Argentina) should be invited to participate in this "food for development" program. And to the extent that competing nations desired to participate, multilateral arrangements could be initiated. This is the way a formal international program under the sponsorship of the United Nations might come into being.

7. But until such time as a world program does come into being, the FAO (or some other UN agency or set of agencies) should be charged with the responsibility, and be provided with the necessary funds, to help recipient countries formulate and execute plans of economic development. This provision would speed development where administrative experience and technical knowhow are most lacking. Further, it would free the United States of the charge of meddling in other countries' internal affairs.

Now, some of you must be saying, "What does this fellow mean by using food to finance economic development? I know what it means to give food away, but how do you use it to finance a five-year plan, or education?" Since my whole argument rests on the use of surplus food supplies to create capital, human and physical, in developing countries, it is imperative that I show how this can be done. Let me do it through two illustrations: (1) the simple case of road building or land clearing in a very poor country, and (2) underwriting a part of the costs of a major plan of development in a country such as India (because of time limitations we must forgo the important but complex case of vocational education and training).

First let us consider the simple case of road building or land clearing in a very poor country. The country involved would plan the project, probably with the technical assistance of the FAO, and mobilize the workers involved, together with their families, into construction camps. The United States would commit itself to providing the kinds and quantities of basic food and fiber products required to feed and clothe the workers and their families for the duration of the project. The United States would further grant the country a loan to permit it to acquire the hard goods required on the project—picks and shovels and some heavy equipment, but not the ultimate in modern earth-moving equipment. The food and clothing costs of the project would probably run to 60 to 70 percent of the total project costs, and these costs we would defray. The foreign country involved would pay the workers a small cash wage in its own currency.

Now let us consider the use of food supplies in underwriting, in part, a national plan of economic development. Assume that India comes to us with a five-year plan, which it may well do, involving the transfer of thousands, possibly millions, of workers first from low-production jobs in agriculture into some kind of vocational training and then into manufacturing and construction jobs. In the early phases of the plan, total output of food would probably decline somewhat; in later phases, the demand for food resulting from the increased productivity of the workers would probably increase more rapidly than agricultural production. Thus, to execute such a plan without serious price inflation the country would need to increase its imports of food supplies for five or ten years. But it is already using its scarce foreign exchange to import the hardware central to the execution of the five-year plan. Here, then, is where we would step in and offer to provide those food supplies at such prices and under such loan conditions as would not impair the financial structure of that developing country. We would agree to supply a given bill of food and fiber goods on the condition that the country did not cut back its normal commercial purchases from us and other national suppliers. This we are doing in sort of an after the fact way with India right now. But we should formalize the procedure and turn it into a forward-operating instrument of policy in the case of India and other responsible national governments (e.g., Mexico and Turkey and perhaps Egypt and Pakistan).

There are shortcomings to this general approach. First, it would cost more than the present program for comparable quantities. Second, many governments in developing countries are not sufficiently strong, or sufficiently responsible, to effectively administer the development plans and projects envisaged. Third, the substitution of development supplies for regular imports will not be stopped in every case, and the demonstration that development supplies are not being substituted for regular imports will never be completely satisfactory to all parties concerned. But to recognize that there would be problems is not to say we should not give these ideas a try. Any course of action, including doing nothing, involves problems.

Many economists criticize me for this stand, arguing as follows: This course of action will perpetuate the farm surplus problem in the United States, and it is an inefficient way to help the developing countries help themselves. It would be better policy, they argue, to move toward freer trade and make dollar grants, or loans, to these countries and thus permit them to purchase what they need for development purposes wherever in the world they are able to get the best deal. To this I would say that they might be right if the economic maladjustments in the world were not so overwhelming. Theirs is an approach that works well when economic sectors and resources are in reasonable adjustment—when we are concerned with interfirm resource adjustments in an economic world that is functioning smoothly. But there are serious maladjustments abroad in the world—some countries are poverty-stricken, others are struggling to keep from falling back into poverty, and still others are cursed with a surplus of agricultural commodities. These kinds of major maladjustments the market does not handle well.

Under the course of action suggested here, two great social complexes, each unstable and unhappy by itself, are made to complement one another. On the one hand we have massive poverty, underemployment, and a revolutionary drive on the part of the peoples involved to improve their worldly lot. On the other, we have great opulence, excess agricultural capacity now taking the form of food and fiber surpluses, and a strange mixture of humanitarianism and fear. The transfer of surplus food and fiber supplies from the United States and their conversion into development supplies in developing countries becomes the policy bridge whereby the pressure of food and fiber supplies on population in the United States is moderated, as is the pressure of population on food and fiber supplies in the developing countries. By this policy bridge we buy the kind of adjustment time required in each social complex; and its construction would constitute political action at its best.

Linking Effective Supply Control to Foreign Demand Expansion

It is one thing to moderate the pressure of food and fiber supplies on domestic population through the sustained use of excess agricultural productive capacity in the United States to finance development in developing countries, and it is quite another thing to keep that pressure from building up again. If we were to run a sustained foreign development program at the level of current surplus disposal operations, *and if the rate of aggregate farm output expansion did not increase*, this course of action would reduce by about half the amount of contraction needed to bring total agricultural output into line with total demand at a level of farm prices slightly above present levels. If, further, we were to run a sustained foreign development program at double the current level of surplus disposal operations, *and if the rate of aggregate farm output expansion did not increase*, this course of action would erase the surplus stock situation in a reasonable time and begin to exert an upward pressure on the level of farm prices. If, still further, we were to run a sustained foreign development program at triple the current level of surplus disposal operations (which is probably all that the non-Communist world could take based on need and a good deal more than it could take based on effective development), *and if the rate of aggregate farm output expansion did not increase*, this course of action would very quickly exert a strong upward pressure on farm prices.

But please note that to each of the situations above, which progress from the realistic to the idealistic, I attached the condition *if the rate of aggregate farm output expansion did not increase* with increases in total demand. But what is to keep the rate of aggregate farm output expansion from increasing? What is to keep the pressure of food and fiber supplies on domestic population from rebuilding? Nothing, so far as I can see, unless farmers accept comprehensive supply control. The total output of food and fiber supplies in the United States is currently outpacing total market demand (domestic and foreign) plus the current level of foreign surplus disposal. And the rate of output expansion would certainly increase under expanded foreign development programs.

For the doubters, let me recall some history. In the late 1930s agriculture was plagued with excess capacity just as it is now. But as the demand for farm products strengthened in 1939–41 with the happy shift from widespread unemployment to full employment, that excess capacity was used up, and both farm prices and farm output began to move up. And when the world's food needs were heaped upon this domestic full employment demand during and following World War II and the total demand for food and fiber products shot forward, the total output of the United States agricultural plant also shot forward. Despite the huge outflow of human resources from agriculture during the war years, 1942–44, and the limited inflow of capital, aggregate output followed closely behind the great expansion in demand of that period. And aggregate output caught up with demand in 1948–49 and has been pressing against it ever since.

I would argue further that the output-increasing potential of American agriculture is greater today than it was in the late 1930s. First, the excess productive capacity that could be thrown into the breach of expanding demand is greater today. Second, total inputs in agricultural production could be easily and readily increased through a slowdown in the rapid rate of out-migration from agriculture. Third, there is no hot war at present to limit the inflow of capital, and farmers' asset positions are generally much stronger today than in the late 1930s. Finally, a broad front of new knowledge currently exists to be adopted in agriculture. In sum, it is my considered opinion that an expansion in total demand resulting from a tripling of present levels of surplus disposal would be matched by an increase in aggregate output within two or three years.

Thus I draw the following conclusions: Any expansion in aggregate demand from any source that might be contemplated now or in the foreseeable future will be quickly followed by an expansion in aggregate output such that the pressure of food and fiber supplies on population is reestablished, with the consequent depressing effects on prices, a buildup of surplus stocks, or both. Hence the receipt of good and stable prices and incomes by farmers generally is dependent on their widespread acceptance of supply control.

I have recently discussed the mechanics of supply control in some detail,[10] and I shall not repeat the discussion here. But we must be clear on terms, and we must share some common ground with respect to general methods, potentialities, and problems of supply control. By supply control I mean the conscious adjustment of supply *to* demand, commodity by commodity, year after year, to yield prices in the market that have already been determined as fair by some responsible agency. And we recognize that numerous avenues of adjustment have been tried, or discussed, in connection with supply control, although most have been conceived as emergency measures and in a halfhearted spirit. But farmers have taken and continue to take some steps in this direction. There are, first, the voluntary efforts of cooperative marketing associations to control marketable supplies. There are, second, the efforts of government to control supplies through the control of one input—usually land, though occasionally the more daring ideas of a planned movement of labor out of agriculture or a tax on the use of a key input (e.g., fertilizer) are considered. There is, third, the use of sales quotas to limit supplies to some market goal.

The first of these efforts has never proved effective except where linked to the third. The second general method has been employed many times in many forms with varying degrees of success. The third general method has been employed with considerable success in sugar, tobacco, and certain specialty crops, but it has not as yet found acceptance among the major problem commodities—not in the great feed-livestock complex, including milk, or in wheat or cotton.

It is my contention, however, that farmers must come to accept supply control of the rigorous type involving the use of sales quotas. This contention is based on two strands of reasoning: the general conclusion reached above that total supplies will quickly adjust upward to match any expansion in demand, and the presumption that supply control based on adjustments in any single input (e.g., land or labor) is blunt and ineffective. Land retirement and the movement of labor out of agriculture are blunt processes that do not lend themselves to fine adjustments in supplies; only the roughest relation can be established between a planned reduction in the employment of land or labor and the output response—and this is not good enough where demand is highly inelastic. And as the state of the art advances, our ability to substitute one factor for another increases; hence we can expect the ready substitution of uncontrolled inputs for the controlled inputs under this type of control, and an ineffective supply adjustment. Trying to adjust supplies through the control of one input is like trying to bail out a leaking boat with a sieve; it is next to impossible.

In the world of effective and comprehensive supply control the composition of demand is all-important; for remember that what we are doing is adjusting supplies *to* demand-at-a-price. In thinking about this control problem it is convenient to break total demand into four components: (1) domestic market demands; (2) special domestic program demands (e.g., school lunches); (3) foreign commercial demands; and (4) special foreign program demands (e.g., to finance economic development). The first component is given to the problem by the level of employment and consumer tastes, the second contains precious little room for expansion, and the third, too, is given to the problem. It is with respect to the fourth component, the special foreign program demands, that real room for decision exists. Decisions made with respect to this fourth component will determine in large measure the size of the initial bite of supply controls on producers for several years to come; they will determine the size of painful downward adjustments required of American agriculture. In this way domestic supply control and foreign surplus disposal are interrelated: the magnitude of the former in the initial stages, at least, depends on the magnitude of the latter; the latter without the former is of limited value to American farmers; and in operation they fit hand in glove.

My task here is completed; the marriage of foreign surplus disposal and comprehensive supply control has been recorded. The mechanics of this marriage union are sketched in the foregoing paragraphs relating the adjustment of supplies *to* demand; the logic of the union has been the work of the whole paper.

2 | Policies for 2002 and Beyond

4 | A FOOD AND AGRICULTURAL POLICY FOR THE TWENTY-FIRST CENTURY

This essay had its genesis in 1998–99 as the prices of the export commodities— wheat, corn, and soybeans—declined sharply and American farmers once again fell on hard times. I wrote it at the suggestion of Mark Ritchie, president of the Institute for Agriculture and Trade Policy, to help farmers and farm leaders understand the economic fix they were in. I argue that American agriculture, although now a part of a global market, is still subject to wide price swings and economic instability. But in that context the old polices of price support and acreage controls are no longer applicable or acceptable. Also, the agribusiness community has metamorphosed into huge monopolistic institutions that farmers must now deal with. I lay out the kind of federal farm policies that must be put in place and made operational if farmers in general are to prosper and family farmers are to survive.

We begin this look to the future by first gaining an understanding of where we came from. We do this with an appraisal of policy efforts in the food and agriculture sector in the twentieth century.

The Twentieth Century Revisited

In the twentieth century America had four basic policy goals for the food and agriculture sector: producing an abundant supply of food, at reasonable prices, for the nation's people; maintaining a prosperous and productive economic climate for the farmer-producers of that food supply; maintaining a family farm type of organization in the production of that food supply;[1] and realizing a high quality of life for all individuals living in rural areas.

The first goal was achieved relatively easily. An abundance of resources available for food production in conjunction with a highly successful scientific and technological development effort made this possible.

The second goal, a prosperous economic climate, met with much less success. Farmers experienced a wonderful economic high during the first two decades of the twentieth century; they fell into a depression in the 1920s; fell into a deeper depression in the 1930s; once again enjoyed economic prosperity in the 1940s; experienced falling prices and economic hard times in the 1950s; and enjoyed moderate prosperity in the late 1960s and early 1970s. Then farm prices fell and hard times returned in the late 1970s; hard times returned again in the middle 1980s; and farmers are once more experiencing an economic depression in the late 1990s. Each downswing is attributed to some specific cause. But why do these specific causes induce "feast or famine" behavior in the food-producing industry—in farming?

They do because the food-producing industry is inherently unstable owing to the highly inelastic aggregate demand for food on the one hand and, in the short-run at least, a highly inelastic aggregate supply of basic food products on the other. Thus any *small* shift in either the aggregate demand for food or the aggregate supply of food products causes a *large* price response up or down depending on the nature of the shift in either demand or supply, with a consequent change in farm incomes. These specific causes (e.g., war, or peace, or great drought, or great technological breakthrough) are always at work shifting either aggregate demand or aggregate supply. Unfortunately, the reality that the food-producing industry is basically unstable is something that most political leaders, farm leaders, farmers themselves, and agricultural economists either don't understand or don't want to understand. The food-producing industry cannot and will not level off at some desirable economic level and stay there. The economic forces of inelastic aggregate demand and inelastic aggregate supply won't let it.

The support programs put in place over the years—price supports, deficiency payments, and acreage controls—lacked the capacity, and in many cases the appropriate design, to cope with the great downward swings in farm prices generated on some occasions by a faltering aggregate demand and on other occasions by a surging aggregate supply. It took the mammoth programs of World War II, and the once in a lifetime increase in the demand for farm products that they created, to pull the farm sector out of the deep depression of the 1930s. In the 1960s farmers and their political leaders rejected the idea of turning the food-producing industry into a government-managed monopoly, a public utility, as a way of coping with the unruly shifts in aggregate demand and aggregate supply. Perhaps at the time that was the proper decision for the nation. But once again in 1999 a faltering demand has pushed the food-producing industry into a depression; this time it is a faltering foreign demand in the brave new global economic world.

Efforts to achieve the third goal (maintaining a family farm organization) have met with even less success. The total number of farms in the United States has declined from 6.5 million in 1935 to 2.05 million in 1999, and most of this huge decline took place among family-type farms. Further, the figure 2.05 million is misleading, since 1.3 million "farms" out of that total are limited resource or residential or retirement farms. That leaves us with about 750,000 real producing farms. Of that total some 575,000 are small to medium-sized family farms grossing less than $250,000 per year, with up to 27 percent of them tied to a marketing or input-providing firm by some kind of contract. These 575,000 small to medium-sized family farms produce some 30 percent of total national production. Of the 750,000 real producing farms, some 163,000 are large to very large family proprietorships grossing over $250,000 per year or are some form of nonfamily farms (corporations or cooperatives run by hired managers), with up to 63 percent of them contracted to a marketing or input-providing firm. These 163,000 large to very large farms produce some 61 percent of the total national product. (For the record, the 1.3 million limited resource or residential or retirement farms produced 9 percent of the total national farm product.)[2]

So after sixty-four years (from 1935 to 1999) and the loss of some 4.4 million farms, most of which were family-type farms, we now have a farm structure comprising 575,000 small to medium-sized family farms struggling to survive in the midst of a farm depression and 163,000 large farms also caught in a sharp price decline, but more often than not linked in some kind of contractual arrangement to a large agribusiness firm. How did this happen? There are a variety of explanations. Certainly some farmers fell by the wayside because they were poor managers. But does that explain a net loss of 4.1 million farms? I think not.

Three explanations seem the most relevant. First, the roller coaster behavior that the farm sector experienced in the twentieth century caught many farmers in exposed financial positions, and as a result they were forced out of business when their product prices fell sharply. Families just getting started in farming and those expanding their operations were especially vulnerable. Second, the farm programs that were put in place supposedly to help the family farmers in fact contributed to the demise of many of them. The support programs provided income or price support in accordance with the number of units a farmer produced—the more the farmer produced, the more price or income support he received. Large producers with the large subsidies from the government thus had their financial position strengthened and could then buy up the assets of smaller, weaker farmers who had received only a small subsidy from the government. The support programs helped the strong become stronger and made the weak weaker—until in fact the weakest dropped out of farming. Third, the rush to adopt modern technology forced many farmers out of business—particularly the smaller ones or those on the financial edge. Much of the new mechanical technology was not size-neutral—large tractor hookups required that their huge costs be spread over large acreages if farmers were to gain economic efficiencies from adopting them. Buying great machines at great costs pushed the adopters in the direction of acquiring more land. And where could they get it? From their smaller neighbors, of course. So another small farmer went out of business to satisfy the needs of a great new machine. And even size-neutral technologies like new and improved seeds put additional financial burdens on the farmer to purchase the new seeds and then the fertilizer and herbicides to go along with them. Thus the costs of putting in a crop of corn skyrocketed in the second half of the twentieth century, causing many a farmer on the financial edge to sell out to his more successful, aggressive neighbor. And by this action we get to the numbers presented above. The total number of farmers declines, and productive resources are concentrated into fewer and fewer farmers' hands.

That is where we stand in 1999. A relatively small number of large to very large producers generate the bulk of our grain, livestock, and fruits and vegetables. Around this core of large producers we find a ring of small to medium-sized family farmers struggling to survive, producing less than a third of the total agricultural product. I say struggling to survive because these farmers are at a disadvantage relative to the large producers with respect to the terms under which they obtain short-term credit and long-term capital. And in this new biotech age, they will be prohibited from obtaining certain of the newer technologies unless they enter into production contracts with the agribusiness firms supplying those technologies. And in all probability the farm depression of the late 1990s will finish off a few more of this increasingly rare species of farmer.

With regard to the fourth goal it is questionable whether the quality of life in rural areas has improved over the past hundred years. Certainly the poorest of the poor still remain out there—in small rural towns, in backwater agricultural areas, and in immigrant labor camps. Federal government efforts to aid these people have been minimal, and even those efforts have been opposed by conservative farm organizations. But probably the most important negative factor has been the decline in the number of family farms and the decimation of the human population in the countryside. As farm families have disappeared, so has the support for small towns and the services they provided—health and educational services, and shopping and repair services. In many farming areas neighborhood activities have simply disappeared with the people.

There are, however, some pluses. Better roads and the automobile have made communication among most rural people easier and more convenient. But the program that did the most to improve the lives of farm people, and certainly the lives of farm women, was the Rural Electrification Administration (REA). When President Franklin Roosevelt initiated this program by executive order in 1935, no more than 5 percent of American farms had electricity; by 1979 some 90 percent did. Electric power took much of the drudgery out of farm work and gave farm people the amenities common to urban living—the radio, lights to read by, hot and cold running water, and refrigeration for food storage.

Improving the quality of life of rural people was an elitist concept from the beginning, pushed by the two Roosevelt presidents—Theodore and Franklin. The idea caught on among their more progressive or liberal followers, but not among the broad body politic. Thus politicians, the ones who vote money to support programs, in one Congress after another throughout the twentieth century provided funding for programs to improve the quality of life in rural areas in a most niggardly fashion.

Looking Ahead to the Twenty-first Century

It is now 1999, a few months before the beginning of the twenty-first century, and farmers are once again experiencing economic hard times. We may properly ask, What are America's goals for the food and agriculture sector in the twenty-first century, and by what policy routes are those goals to be achieved? America's goals for the new century appear to be much the same, with some modification, as those of the past century: producing a *healthful*, abundant supply of food at reasonable prices for the nation's people; maintaining a prosperous and productive economic climate for the *commercial* farmer-producers of that food supply (recall that over half of the units now defined as farms produce little or nothing for sale); protecting the remaining 575,000 small to medium-sized family farm units from disappearing from the face of the earth; and realizing a high quality of life for all *humans* living in rural areas, *together with a vibrant physical environment*.

The goals of American society may not have changed greatly over the past hundred years, but the conditions under which its food supply is produced and distributed certainly have. In the last several decades of the twentieth century the productive resources of the food-producing industry have been concentrated in a relatively small number of large, industrialized producing units (on some 163,000 units—I hesitate to call them farms); inputs on these producing units are strictly controlled with the aid of computer technology, which in certain livestock operations approach factory conditions; in many, if not most, cases these local producing units are tied through contracts to either an input-providing agribusiness firm or a processing-marketing firm where the key business decisions are made with respect to the operations of the local producing units; and all of this is taking place in a rapidly changing and often unpredictable global economy. Farming is not farming anymore; food production is an industrialized business in which operating decisions are commonly made far removed from the local producing units.

TROUBLESOME AREAS

In this context of industrialization and globalization, farm prices and incomes plummeted in 1998–99. Some argue that the increased monopolization of the input supply side of farming, together with the product marketing side, caused these disastrously low farm prices. Others argue that the sharp decline in farm prices in 1998–99 resulted from a contraction in foreign demand. As will soon be evident, I hold to the latter view. *However*, I am strongly of the opinion that with the increased use of such business practices as contracting, patent rights, and financial controls, the food production sector—farming—is being converted into a poorly understood area of monopolistic competition. And continued developments in this direction could have fearful consequences for both farmers and consumers.

Developments in this area are crying out for explanatory analyses comparable to those of Edward Chamberlin and Joan Robinson in the 1930s. Lacking those analyses, in my policy recommendations I argue for an active, innovative antimonopoly division in the Justice Department to explore and deal with the monopolistic problems developing in commercial agriculture. The standard legal concepts of monopolistic action, collusion, or connivance among parties may be a part of the emerging monopolistic problems, but they are not the essence. *Bigness is the problem*, and the power that bigness brings with it. Introduce a giant corporation providing a commonly used farm input into a local farming community, and it will have an advantage in every transaction or activity it enters into. These may range from fixing the terms of a sale to, or a contract with, a local farmer, to obtaining from the local government a tax-free site for its plant, to beating up its local competitor. A giant modern corporation operating in a local farm community is like a bull elephant in a china shop. The power of the giant overwhelms and shatters the local establishment.

In commercial agriculture we have two different developments that constitute problems for family farmers because the outcome of those developments will determine the farmers' fate. The first is the continuing struggle, often minimized by economists, over who controls the use of productive resources on farms—the individual family farmers or some powerful outside purveyor of a key input (e.g., a genetically modified seed)—whose outcome will determine whether there are any independent family farmers left in the year 2010. The second is the immediate life or death price-income crisis, whose outcome will determine which family farmers are still in business at the end of this year or the next. Two different problems, with similar consequences.

In connection with the second problem, it is sometimes suggested that a return to the old commodity programs with their price and income supports in conjunction with acreage controls would be an appropriate means for assisting food producers, large and small, in the difficult economic times in which they find themselves in 1998–99. Such a policy action would be a major mistake. This is true for a number of reasons: Under that program arrangement, most of price and income benefits would, as in the past, go to the large producers who need them the least, and only driblets would reach the small to medium-sized producers who need them most. The large industrialized producers, geared to a high level of production, are not likely to accept production controls, but supporting farm product prices or incomes above market levels, even in depressed periods, without effective production controls, can lead to only one result—burdensome stocks in government hands and unacceptably large program costs. And finally, imposing effective production controls to reduce available supplies in a country like the United States as a means of reducing global supplies would be self-defeating—competing nations would simply increase output, and total world supplies, after a brief period of adjustment, would remain the same or possibly increase as the United States' share of the world market declined. The old commodity programs have no place in the modern global economy. But there are problems ahead for an inherently unstable food and agriculture sector operating in an unpredictable global economy.

In the 1980s it was popular to talk about the high elasticity of export demand for, say, soybeans (how high was much debated); foreign purchasers of American soybeans would increase their takings of American soybeans, or soybean meal, at a reduced price first by substituting American soybeans, or soybean meal, for other high-protein feed supplements and second by substituting American soybeans, or soybean meal, for soybeans from some third country. And the possibility of this high elasticity of export demand for other American agricultural commodities was viewed by some farm leaders and many farmers as the magic solution to the "farm problem." And this view of an expanding export market as a solution to the "farm problem" continues down to the present day. But this game cannot go on indefinitely in a finite trading world. Everyone—Americans, Canadians, Brazilians, Frenchmen—cannot expand sales of their product by substituting it for that of someone else (who happens to be trying to do the same thing) unless, of course, the total market has no limits. But events of 1998 and 1999 have made it clear that the global market does have limits; in fact the global market for food products contracted in those years. The global market for food products may be wide, and there may be lots of room for individuals to maneuver within that market, but it does have limits.

I submit that the way to look at the global market for food products is the way we looked at the aggregate demand for food in the United States in the 1950s. The global market for food products is a closed system with a given population distribution, a given income distribution, a given set of national laws and rules, and a given set of human tastes and preferences. Taking account of these givens, there exists at any point in time a *global demand for food products*; this concept may be difficult to measure, but it is real, and it is *highly inelastic*. It is highly inelastic because the stomach of each member of the world population is highly inelastic. The human stomach craves roughly the same amounts of the same kinds of foods day after day.

The concept of the global demand for food products is illustrated in the accompanying figure by the line DD_{97}; it has an elasticity of approximately 0.25 at the price level 100.[3] Given the demand curve DD_{97}, a global production of 100 yields a price solution of 100, which we take to represent the situation in 1997 (point A in the figure).[4] Now, let us assume a contraction in demand of 5 percent at the price level 100, illustrated by the new demand curve DD_{98} in the figure, which I believe is in line with what actually happened in 1998. Given this contraction in global demand, the production of 100 can no longer clear the market at price level 100; the price level must fall to 80 to clear the market, or a decline in the price level of 20 percent in response to a 5 percent contraction in demand. But this is not the end of the story. Global production is increasing at about 2 percent per year during this period. So the price level must fall to 72, or by about 28 percent, to clear the market in 1998 (point C in the figure). Again, I believe this is consistent with what actually happened in 1998.

The global market for food products moves along in 1999. We assume world production continues to increase at about 2 percent per year, as illustrated by SS_{99} in the figure, and global demand recovers by about 2 percent in 1999, as illustrated by DD_{99}. These develop-

An idealized global market for food products, 1997–99

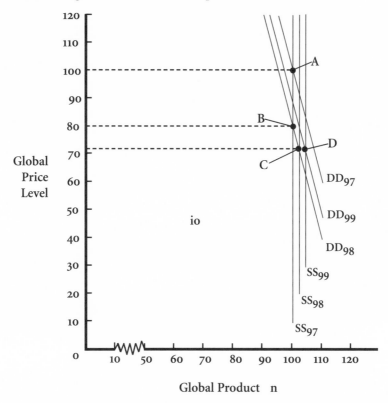

ments yield a market clearing price level in 1999 with little or no change from the 1998 level (point D in the above figure). The persistent increase in world production of food products in recent years in combination with the erratic shifts in the severely inelastic global demand for those products has led to disastrously low prices for those products and to business failure for many American farmers. It should be noted that all food product prices will not rise or fall in lockstep; the major products most directly affected by developments in the global economy will lead the way, and their pricing behavior will spread to other products through the ubiquitous substitution process in the global market system.

As the previous discussion suggests, both the global demand for food products and the aggregate supply of those products will be expanding through time as human populations and national economies continue to grow. But the demand and supply relations are unlikely to grow at the same pace; the determinants of these two relations are unrelated. Population growth and income growth are the principal determinants of the demand relation; investments in research, new technologies, and various capital items (e.g., irrigation works), along with climate changes, are the leading determinants of the supply relation. Thus it seems likely that on some occasions aggregate global supplies of food products will push ahead of aggregate demand, and farm prices will fall sharply; on other occasions aggregate demand will push ahead of aggregate supply, and farm prices will shoot skyward. Whenever either happens, which in my view will be much of the time, we must expect a change in the food product price level. And often those changes will be dramatic.

To the uninitiated those price level changes in the food product area may appear out of the blue like acts of God, but they are not random acts. They are the results of the interaction of a severely inelastic global demand for food products and an unpredictable aggregate supply of those products in the context of a global economy. We will continue to experience these severe price level changes in the food products areas, like the one that is occurring in 1998–99, so long as humans dictate the nature and the slope of the aggregate demand curve and we operate in a relatively free trade global economy. Thus it behooves food producers large and small to recognize that they are operating in a high-risk industry. Domestic policies can mitigate the impacts of these price risks; they cannot eliminate them.

In this connection, history tells us that boom times are the dangerous times for farmer-producers; it is in such periods that they tend to get carried away, expanding their operations with generous applications of credit, forgetting that boom times have always been followed by hard times—times in which those who are overextended creditwise go broke. To repeat, agriculture is a high-risk, unpredictably unstable industry. And those farmer-producers who did not learn that lesson in 1998–99 are not likely to be among the survivors ten years hence.

A Food and Agriculture Policy for the Twenty-first Century

Below I outline, with brief comments, a set of policies, some new, some old, to help Americans achieve their goals for food, agriculture, and rural areas in the twenty-first century. This comprehensive set of policies represents what I believe are the necessary actions the federal government must take if consumers of food, agribusiness firms, and farm producers are to survive and prosper in the rapidly changing world of technology, business organization, weather and climate, and the global economy in the decades immediately ahead.

I. Policies aimed at achieving an abundant, healthful supply of food
   A. Maintain the existing food assistance programs (food stamps, school lunches, special programs for feeding nursing mothers, pre-school-age children, and the aged and infirm), but greatly improve the service provided, and the quality of meals and food supplements provided.
   B. Greatly strengthen the Food and Drug Administration so that it can properly inspect all imported foods and all food processed and handled in the United States, with the authority to ban and destroy all items that do not meet the standards of the agency.
   C. Maintain a strong research program in the U.S. Department of Agriculture and the colleges of agriculture across the nation in all aspects of food production, processing, distribution, and human nutrition.

II. Policies aimed at achieving a prosperous economic climate

    A. Establish a Food Production Refinancing Agency, with adequate capital, to help any food-producing units (small, medium, or large) refinance at subsidized interest rates or relaxed payment schedules in periods of falling prices and income. The purpose here is to help firms restructure their debt loads so that interest charges and other loan terms are consistent with the new lower levels of product prices. In addition, this agency should provide guidelines to farm lending institutions (both public and private) governing the extension of production credit to farmer-producers—especially in boom times.

    B. Establish a special disaster program focusing on the problems of food producers to help them reestablish or relocate their producing units following a flood or drought or plague or any other natural disaster. Such a program will be indispensable to the food and agriculture system, hence the nation, should weather patterns become increasingly violent with global warming.

    C. Establish a grain reserve program large enough to enable it to help moderate large swings in world grain prices and to act as the agent of the federal government in providing relief supplies to areas of great need and famine around the world. In its stabilizing function the reserve program would, of course, acquire stocks in periods of falling prices and dispense them in periods of rising prices, but it should *not* attempt to support grain prices significantly above world market levels. Such a grain reserve program opens up numerous farm storage possibilities from which producers could earn storage payment income.

    D. Transform the Office of the Trade Negotiator into a Trade Policy Action Agency with two basic responsibilities: first, it should work to open markets to U.S. products abroad, with emphasis on agricultural products; second, it should work with other nations and international agencies to establish rules of conduct, conventions, and possibly treaties to protect the physical environment and safeguard the quality of life of workers in the conduct of international trade.

III. Policies aimed at maintaining a family farm organization

    A. Create a special unit in the Department of Justice to investigate monopolistic actions in the food production and distribution system and to prosecute firms taking such actions where they are deemed harmful to the efficient and *equitable* operation of the food and agriculture sector of the national economy. This new unit should have special expertise in the new world of biotechnology, contract farming, and firm behavior under conditions of monopolistic competition and oligopoly, and it must recognize that bigness alone is an important aspect of monopolistic action.

    B. Make an annual cash subsidy ranging from $15,000 to $25,000 (depending on size of operation) to each of the 575,000 family farm units with gross incomes below $250,000;[5] this subsidy is made to help these small to medium-sized units stay in business in competition with the larger industrialized units, where they are at a disadvantage in obtaining credit and capital as well as in certain purchasing and selling transactions where product volume and business connections play important roles.

    C. As national programs for sustainable agriculture develop, these surviving family farmers should be brought into the programs as key players. Here is an opportunity to weld together two programs that complement one another and gain strength in the process. The possibilities here are limitless for introducing innovative production, conservation, and management practices. A strong extension program must be a part of this policy package to help these small to medium-sized farmers swim against the tide, namely, that bigger is better.

IV. Policies aimed at achieving high quality of life

    A. Create a special unit, knowledgeable about agricultural production practices, within the Environmental Protection Agency to monitor and regulate the use of nonfarm-produced inputs. This organization should have the authority to ban the use of such inputs where it is established that they harm the environment or the humans living in it.

    B. Establish a federal program to monitor and regulate factory-type operations in the production and processing of poultry, beef, pork, and dairy products. The agency administering this program should have the authority to set standards and enforce those standards in

        1. the confinement of birds and other animals,

        2. the location of such factory type operations

        3. the disposal of the wastes of the animals confined in such factories

        4. working conditions of human laborers in such factories.

    C. Support an adequate income and safe working conditions for agricultural workers by

        1. creating a special unit within the United States Department of Agriculture to examine laws and regulations affecting farm workers and to enforce current laws. Employers should be held accountable to safety standards established by the

Occupational Safety and Health Administration
and the wage, overtime, and insurance provisions
of national labor law
2. affording agricultural workers protection
guaranteed most other workers under the
National Labor Relations Board.
D. Maintain the Conservation Reserve Program, but
strengthen its conservation provisions by
1. limiting rental contracts to low-grade cropland
2. extending the duration of rental contracts to very
long periods (e.g., thirty to forty years)
3. purchasing the land involved where it can be
phased into the ecological area program outlined
in the next recommendation.
E. Create a new federal agency with the responsibility,
authority, and requisite funding, working with private
organizations and state and local authorities, to build
ecological areas for different species of plants and
wildlife that are sustainable. This should be an ongoing
operation that builds these ecologically sustainable
areas from existing federal lands and the purchase of
private lands.

The policy agenda laid out above is an ambitious one. It would take a
strong president like Franklin Roosevelt or Lyndon Johnson to get it
passed into law. But if it was passed in its entirety, was adequately
funded, and was put in operation by leaders who believed in it, it
would make rural America a pleasant place in which to work and live.
But most important, it would give the food and agriculture sector the
program tools to ride out, though not without some pain, the economic bumps it will surely encounter in the years ahead as a part of
the global economy.

But even if, as is more likely, only pieces of it survive in the legisla-
tive battles ahead, the full agenda, as presented here, provides a direc-

tional guide to the development of an equitable and effective food and agricultural policy in the years to come. The downturn in product prices in the food and agriculture sector in 1998–99 is not the last this sector will experience. And changing weather patterns as a part of global warming will certainly test the skill and fortitude of food producers in the years to come. The twenty-first century may bring great opportunities for some, but it is likely to bring great problems to both family farmers and the large industrialized producing units.

5 | THE EXPORT SOLUTION

I wrote this piece as the basis of a news release by the University of Minnesota Extension Service in a series dealing with current problems confronting farmers. I meant it to counter the loose talk making the rounds in the upper Midwest, claiming that all that is needed to restore prosperity to the farm economy is a small increase in exports. How this small increase in exports was to be achieved was blissfully ignored. In a larger sense, I try in this short piece to educate farmers about the perils of relying too heavily on the export market.

I begin this discussion by recognizing that a strong export market is vital to the well-being of the American farm economy. In 1996, a good year for the American farm economy, farm commodity exports accounted for approximately 30 percent of the total value of farm production. In 1999, a poor year, farm exports accounted for only some 26 percent of the total value of farm production.

These numbers tell us that a major portion of U.S. agricultural production is exported annually. But they do not tell us why farm product exports declined in importance to American farm producers, in value terms, during 1996–99. The picture is further confused by the fact that the physical quantities of wheat, corn, red meat, and poultry products exported increased modestly over the period 1996–97 to 2000–2001, while the physical quantity of soybeans exported first declined and then rose. What happened, of course, is that the global demand for wheat, corn, soybeans, and pork contracted, and the prices at which those products could move into use or consumption plummeted. And with those sharp price declines, the value of U.S. farm product exports declined.

Some farmers and their leaders, not understanding how markets work in the global economy, particularly the role of demand in determining prices, and seeing their gross returns falling from declining product prices at home, are inclined to reason, "If we could send more of our products abroad, that would take the downward pressure off our domestic prices." But such reasoning ignores the fact that domestic prices of the basic commodities—wheat, corn, and soybeans—are determined in the global market, and all things being equal, shipping more of those commodities abroad would drive down prices in the global market and in turn lower our own domestic prices. Local markets for those basic commodities are part of the global market; there is no great unseen foreign market onto which we can dump surplus American farm products with no effect on our domestic markets.

Those who would solve our domestic problems of too much production by shipping surplus products to some faraway foreign market need a better understanding of the operation of the global market, of which we are now an integral part. We can visualize a global demand for each of the major farm commodities (wheat, corn, and soybeans) made operational by a variety of buyers (traders, speculators, processors, government agencies) around the world. The price elasticity of the *global demand relation* for each of these major farm commodities probably ranges from a high of -0.7 for soybeans and soybean meal to a low of -0.3 for wheat, and to -0.5 for corn (this means for corn, for example, that with a 10 percent decline in its world price only an additional 5 percent of it would be taken). Each of these demand relations is a derived demand—derived from the demands of consumers for the finished products of the commodity. These ultimate consumer demands are the product of four basic determinants: the size of the human population; the income distribution of that population; the tastes and preferences of that human population; and the rules and laws of each nation governing the handling and distribution of the finished products. The content of these determinants for any point in time establishes the demand relation (i.e., the number of units demanded for each price, hence the measure of price elasticity) for each finished product in the global market. The summation of the demands for all the finished products derived from a given farm commodity (e.g., corn), *as interpreted by the various traders*, yields the operational demand in the global market for the commodity in question.

Opposing each farm commodity demand relation is, of course, a concept of the global supply of each commodity. It is conventional to present this concept of supply as a function of price, as in the case of the demand relation. But I will not do that here because I do not believe that the glob of global production of each commodity (e.g., corn) can be stated as a simple function of price. The glob of supply that becomes available in the global market at any time is a function of many economic and physical variables, some of which are completely unpredictable (e.g., the weather). In my view the magnitude of the glob of supply of a given commodity in the global market is a function of past and present commodity price movements, past research and technological developments, past investment decisions (both micro and macro), past monetary and credit policies, national government programs subsidizing the production and the export of the commodity, and last but not least, past and current growing conditions and weather. What turns up as the global supply of a given commodity at a point in time is thus largely unpredictable.

This unpredictable global supply of a given commodity interacts with the global demand to yield a global price level for the commodity. I say price level because specific prices in specific locations will vary because of variable transportation costs, and perhaps because of monopolistic chicanery by certain traders. If that global demand for the commodity is increasing relative to the glob of supply owing to population growth and rising per capita incomes, the price level of the commodity will be rising and farmers will be happy. But if that global demand for the commodity is decreasing relative to the glob of supply, the price level of the commodity will be falling and farmers will be unhappy. The price level of the commodity in question (e.g., corn) is determined and redetermined in a continuing process of haggling and bargaining by buyers and sellers in that worldwide market.

What American farmers see is not the operation of the global market, but only their own falling commodity prices, as was the case in 1998–99. Thus they and some of their leaders reason, If we could strengthen our local markets by expanding the exports, all would be well. But this is not easily done. The export market is not some infinitely expanding space, like the universe, into which some federal agency can simply shoot surplus American farm commodities. The global market for farm food commodities *is a closed system, in which the determinants discussed above establish the boundaries of the demand relations for those commodities*. Over time population growth and rising per capita incomes have been the principal sources of the expansion in the global demand for such American-produced commodities as wheat, feed grains, soybeans and soybean meal, poultry products, and many specialty products. But when population growth levels off, as it has in Europe and Japan, and countries of the Pacific Rim experience an economic recession, there is nothing that American policymakers can do to instantaneously increase demand for America's farm commodity by causing populations to grow or by reviving per capita incomes in economies experiencing an economic downturn.

But, you say, the United States should engage in sales promotion and advertising to entice foreign consumers to buy American food products—that is, lure them away from the products of their own domestic producers, or from those of export competitors of the United States. Here you have in mind changing the tastes and preferences of foreign consumers in favor of American food products through sales promotion and advertising. This we have done in the past, and sometimes it works. It works for certain specialty products like a brand of olives, or a brand and type of wine, or possibly even poultry products. It works where the products are special and where they can be differentiated with the use of brands. The trick is to get foreign consumers tied through advertising to a particular American brand. But how do you tie a foreign consumer to a particular grade of American wheat, when a bushel of wheat of the same grade and type is the same whether it is produced in Australia or the United States? It is almost impossible to expand the sales of undifferentiated products through advertising. So we see that sales promotion and advertising have a place in expanding exports of certain American farm products, but it is a limited place.

Since the days of the New Deal, the United States has sought to open foreign markets by negotiating trade agreements. In these negotiations the United States has always, except when dealing with Communist nations such as the Soviet Union, pursued a multilateral approach in which the agreements included the "most favored nation" provision. This provision means that any trade advantage granted the United States by the foreign nation in the agreement becomes available to all other trading nations. Such agreements have been important over the long run in providing an opportunity for U.S. producers of farm products to increase their exports to these countries as they developed economically. But such agreements made in the philosophy of free trade do not give any special trading advantage to the United States or involve a commitment by the foreign nation to import a specific quantity of U.S. produce. And since both Democratic Party leaders and the present Republican administration are ardent supporters of a free trade philosophy, it is unlikely that any bilateral trade agreements will be negotiated in the near future. We will continue to negotiate trade agreements with multilateral implications.

This avenue of export expansion for farm food products thus provides the *opportunity* to increase the sales of U.S. products to the countries involved as they develop economically and experience rising incomes. But such agreements do not necessarily provide any boost in exports to the countries involved in the short run. They do not contain commitments by the foreign country to take additional amounts of a particular product in a specific time period.

There is one other important way for a country's trading agents to increase the exports of its farm food products. Traders, public or private, can simply lower their selling price and take sales away from traders in countries who either are unwilling to lower their selling prices or are tardy in doing so. This, as you see, does not involve any expansion in the global demand for the product; it involves moving down the existing demand curve for the product.

This is an ideal solution for a nation that is a low-cost producer. And it is probably how Brazil has increased its share of global trade in soybeans and soybean meal over the past decade. Brazil, a low-cost producer of soybeans because of low land and labor costs, can afford to shade its selling price below quoted world prices and take sales away from nations unwilling to do so. This is price competition, and there is nothing irrational about it in a free market setting where *the export demand* confronting the producers is elastic and if their cost structures permit it.[1]

But for the United States it is a terrible solution. The world prices for products like soybeans and corn, translated into U.S. prices, are already below the average costs of production for most U.S. producers of these commodities. Now, to expand your sales by selling more at a still lower price is no way to get well financially and to stay in business. Because of high land and labor costs relative to Latin America, U.S. producers of corn and soybeans are in no position to get into a competitive price war with producers in Latin America. Sliding down the global demand curve for soybeans, pricewise, can have no appeal for U.S. producers.

To this point I have discussed the policy proposition of expanding exports in terms of individual commodities. But what about the situation in the late 1990s when the prices of all the leading export crops fell, as well as the prices of important livestock products? Proponents of the policy solution of expanding exports are likely to argue that this is the very situation in which it is imperative to expand exports. They say, "This is the ideal solution for this kind of a problem." And it would be, if the policymakers could pull it off. But the point of this essay is that they can't.

The aggregate global demand for farm food products weakened in the late 1990s as the result of a financial crisis in Pacific Rim countries and a slumping economy in Europe. Because of the severe inelasticity of the aggregate global demand for farm food products, a modest decline in that demand caused the prices of those products to fall drastically. Given this inelasticity of aggregate demand, the worst conceivable policy action by exporting nations would be to offer to sell more corn, or wheat, or soybeans at lower prices. To move additional quantities at lower prices in this situation would require that the prices of the products had to fall further and drastically.

If the prices of most farm food products fell because the global demand for them contracted, as was the case in 1998–2000, how could any rational person expect the exports of those commodities to increase through an increase in demand? When the per capita incomes of consumers in importing countries are stagnant or falling, they are not going to increase their purchase of finished products derived from American farm food products because American farmers, or their policymakers, want them to. Imports of American farm food products will increase again only as those importing countries pull out of their economic slump and consumer incomes begin to rise.

In conclusion, for the farm sector of the American economy to remain healthy and viable, it must year after year export 30 percent or more of its total production. To make this a reality the U.S. government must pursue policies designed to contribute to the effective and prosperous operation of the global economy, continue its policy of opening up markets through multilateral trade agreements, and continue to engage in sales promotion for those specialty commodities where that kind of action makes sense. But when the American farm economy runs into price and income troubles, as it will from time to time, efforts to combat those troubles by a quick expansion in exports will in most cases meet with failure (and they most certainly will if the proponents of such a policy have in mind expanding sales by engaging in price competition, which means sliding down an inelastic global demand relation). The global demand for American farm food products cannot be manipulated at the beck and call of American policymakers. It is formed and moved by a set of determinants beyond their control. Fantasizing about solving the price and income problems of American farmers through instantaneous global demand expansion is like fantasizing about winning the Power-Ball lottery. The chances of success are about the same.

6 | SAVING THE FAMILY FARM

The Case for Government Intervention

In recent years I have been asked many times both by farmers and by city people, "What are we going to do to save the family farm?" And I have been told many times by agricultural professionals with whom I associate, "The family farm is dead. Stop worrying about that anachronism." Partly to answer the question, asked in all seriousness, and partly to rebuff some of my cynical associates, I began thinking seriously once again about how to save the family farm. I put those ideas on paper under the title above. As this book took shape in my mind, the ideas developed there became this chapter, which serves as an introduction to the final chapter.

What Is a Family Farm?

Some agriculture experts say the family farm is dead. There is some truth to that: the number of family farms has declined in every decade of the past fifty years, including the 1990s. But as we shall see, they are not all gone. And I argue that we should take positive action to save those that remain.

Right off, I can hear someone saying, What does he mean by a family farm? I define a family farm as a type of organization in which the family makes all the important operating and investment decisions, owns a significant portion of the assets, and provides a significant amount of the labor required.[1] Of course the primary occupation of the head of this farm family is farming. Can this definition be made operational with respect to the size and number of working family farms? Unfortunately, that is not easily done from data available to us. We can, however, deduce some ideas about the size and number of family farms from the following table.

Structure of U.S. Farms: USDA Typology, 1999

	Small family farms					Larger family farms			
	Limited-resource	Retirement	Residential/ lifestyle	Farming/ lower sales	Farming/ higher sales	Large	Very large	Nonfamily	All farms
All farms (number)	195,572	304,293	811,752	396,698	178,210	79,240	45,804	37,816	2,049,384
All farms (percentage)	9.5	14.8	39.6	19.4	8.7	3.9	2.2	1.8	100.0
Value of production:									
Total ($ million)	1,615.5	4,378.2	13,126.7	19,971.5	35,249.7	30,230.7	59,582.5	27,569.3	191,724.0
Average value of production per farm ($)	8,244	14,401	16,186	50,307	198,033	382,670	1,295,282	725,500	93,569
Percentage of total U.S. production	1	2	7	10	18	16	31	14	99
Percentage of farm type with production and/or marketing contracts	2.9	6.7	3.9	11.6	27.2	47.2	62.9	23.1	11.1
Percentage of value of production under contract	8.5	12.4	13.4	23.4	19.4	27.9	44.3	40.1	31.2

Source: Agricultural Outlook, ERS, USDA, Jan.–Feb., 1999, pp. 6, 7.
Small family farms:
  Limited-resource: Operator household income under $20,000, farm assets under $150,000, and gross farm sales under $100,000.
  Retirement: Operator's primary occupation is retired.
  Residential/lifestyle: Operator's primary occupation is "other"—neither farming nor retired.
  Farming/lower sales: Operator's primary occupation is farming, and gross farm sales are under $100,000.
  Farming/higher sales: Operator's primary occupation is farming, and gross farm sales are $100,000–$249,000.
Larger family farms:
  Large: Gross farm sales $250,000–$499,999.
  Very Large: Gross farm sales $500,000 or more.
Nonfamily farms:
  Nonfamily corporations or cooperatives, and farms run by hired managers.

In the table we have a description of the structure of farms in the United States in 1999, as presented by the U.S. Department of Agriculture. From it we can deduce some important, as well as confusing, aspects of that structure. For example, there are 1,311,617 small units that the USDA calls "small family farms." They constitute almost 64 percent of the total number of farms, but they produce only 10 percent of total farm production. At the other end of the structural spectrum are 83,620 units that the USDA describes as "very large family farms" and "nonfamily farms," which amount to some 4 percent of the total number of farms and produce 45 percent of total farm production. Note also that the USDA calls every farm with a family living on it a family farm, whether it is a low-production unit with a gross of $8,244 a year or a family proprietorship grossing $1,295,282 This is a pretty loose concept of a family farm.

The 575,000 farms falling under the categories "farming/lower sales" and "farming/higher sales" in table 1 seem to fit my definition of a family farm. The operator's primary occupation is farming, and the families could provide a significant amount of the labor required on farms of this size. A question arises, however, with respect to the contracting by some of these farms: Is the independent decision making of these families lost by such contracting practices?

Next, what about the 79,240 farms under the category "larger family farms," with average annual gross sales of $382,670? Can the typical farm in this category meet the conditions of my definition? The question here is whether these family proprietors actually live on the farms and provide a significant amount of the labor required. I don't know the answer, but I would guess that in some cases they do and in others they don't.

Let us assume that half of the farms in this "larger family farm" category fit my definition of a family farm, and that three-quarters of the farms in the previous two categories ("farming/lower sales" and "farming/higher sales") satisfy the conditions of my definition. I also assume that none of the farms in the category "very large family farms" do so. That yields a rough estimate of 471,000 working family farms in the United States in 1999, which account for approximately 29 percent of total farm production. That number is a far cry from the 6 million farms that existed in 1939, most of them family farms. But those 471,000, I argue, are worth saving for two reasons. First, they protect our rural countryside from making the Tyson Chicken syndrome universal. Second, they can help prevent our food and agricultural system from becoming a thoroughgoing monopoly.

Two Problems for Family Farmers

What is it that we must save the working family farm from? Two developments have long adversely affected many, if not all, farmers: first, price and income instability (unpredictable and dramatic price and income movements), and second, technological developments and business practices that have led to bigger and bigger farming units and oligopolistic structures in agribusiness. Of course, each farmer-producer will be plagued by his own particular set of problems that he may or may not be able to overcome, but I refer here to those overarching problems that come to bear on all farmers.

PRICE AND INCOME INSTABILITY

Price and income instability was the curse of all farmers, but particularly family farmers, in the 1900s. The upside, of course, presents no problems. But if you are in a business where the prices of the products you sell drop out of sight on occasion but the prices of the inputs you buy hold steady, and if your mortgage payments are fixed, as are the payments on your new combine, then you can quickly move from a positive net income position to a negative one. One day you are financially well off, the next day you can't pay your grocery bills. That is the roller-coaster ride that family farmers experienced throughout the 1900s and are suffering once again at the dawn of the twenty-first century.

Let us take a brief look at the price and income ride that farmers experienced in the 1900s. After twenty glorious years of rising farm product prices, 1900–1920, farmers went through almost twenty years of low prices and disastrous incomes from 1921 to 1940; farm prices and incomes fell sharply once again in the 1950s and early 1960s; after a price boom in the early 1970s, farm prices fell again in the late 1970s, rose in the early 1980s, and plummeted again in the middle 1980s. Farm prices and incomes did relatively well in the early and middle 1990s but fell sharply in 1998 and have remained low since then. In each of those price and income downturns, another batch of farmers went out of business—typically, those who entered the downturn in a weak financial position. But their land did not go out of production; it was incorporated into the operations of bigger farmers in stronger financial positions and probably farmed even more intensely. After each downturn there were fewer family farmers left on the land but more very large units, producing a larger share of the total national production. Farming has increasingly become big business, financed by big city banks and corporate enterprises.

Farming is an inherently unstable industry, and we can understand why if we comprehend the nature of the market in which the major commodities are bought and sold. Those commodities are wheat, rice, feed grains (e.g., corn, oats, barley), oil seeds (e.g., soybeans), and sugar. Those commodities are now traded in a global market, and their prices are determined by thousands of transactions in that market where in each trader is exercising his or her best judgment of what the true global demand and true global supply of that commodity may be. From these myriad transactions there results for any point in time a price—the world price for the commodity in question. That world price, say for corn, is "discovered" by traders in terms of quotable prices on the trading floor of the Chicago Board of Trade and converted into actual sales prices in the commodity surplus area of the ports on the Gulf of Mexico. The price of a specific grade of corn at, say, Fort Dodge, Iowa, is based on the sales price of that grade of corn at the Gulf ports minus the transportation and handling charges from Fort Dodge to the Gulf. Other institutions and possibly other surplus export areas will be involved for wheat and soybeans.

We are back now to the forces of supply and demand. Let us focus first on demand. The global demand for a basic commodity (for example, corn) reflects the judgments of the many traders in the global market as to how much of this commodity will be used to produce food products to be consumed in turn by human beings. So we are back to the human consumer who, at any level of income, seeks to eat about the same amount of the same foods each day, which in turn converts to an inelastic demand for food products. This information threads its way back to traders in the global market for the basic commodities and has the effect of converting traders' demands for these basic commodities into inelastic demands. In nontechnical terms, this means that if there is an oversupply of the commodity in the market, prices must fall a long way to move that oversupply into some use (this is what happened in the global market in 1998–99). And, of course, when there is a supply shortfall in the market, prices for the commodities shoot through the roof.

But the global demand for these basic farm commodities is not fixed in time. Typically it expands slowly, powered by population growth and rising per capita incomes. (With rising income, consumers tend to eat more animal products and less cereals and potatoes.) On occasion, however, this global demand will stop expanding, or even contract, if important or key economies run into a recession or depression. This is what happened to the Pacific Rim countries in the 1990s; global demand contracted modestly, creating an oversupply, and prices fell precipitously. The picture to be gained is thus one of a global demand for these basic commodities slowly expanding through time, but occasionally coming to a halt or even contracting somewhat.

Information about the production and supplies of the basic commodities in the global market is more solid than information about the demand for those commodities. But predicting the supplies available for a given commodity (say, corn) in the next year is more chancy. That is because the production of those commodities is a function of numerous variables: past and present price movements, past research and technological developments, past investment decisions (both micro and macro), past and present government programs subsidizing the production and export of a given commodity, and last but not least, past and current growing conditions and weather. This unpredictable global supply interacts with the global demand to yield the global price level for the commodity. If the global demand is increasing relative to the supply, the world price level of the commodity, and specific prices in specific locations, will be rising, and farmers will be happy. But if the global demand for the commodity is decreasing relative to the supply, the world price level will be falling, and farmers will be unhappy.

This is a dynamic process. Both the global demand for these basic commodities and the supply of them will be expanding through time as population continues to grow and national economies develop. *But they are unlikely to grow at the same pace*; the determinants of these two relations are unrelated. Population growth and income growth are the principal determinants of the demand relation; investments in research, new technologies, and various capital items (e.g., irrigation works) along with climate changes are the leading determinants of the supply relation. Thus it seems likely that on some occasions the global supplies of the basic commodities will push ahead of the global demand for them; then prices will fall sharply because of the inelasticity of the global demand. On other occasions the global demand for those commodities will push ahead of supplies, and their prices will shoot skyward. The former was the more common occurrence in the twentieth century. What the twenty-first century holds for us remains to be seen.

But we do know that we will continue to experience severe price level changes for the basic farm commodities so long as humans dictate the slope and elasticity of the global demand curves for those commodities, the production and supplies of these commodities remain unpredictable and we operate in a relatively free trade global economy. Thus it behooves producers of these commodities, large and small, to recognize that they are operating in a high-risk industry.

Domestic policies can mitigate the effects of these sharp price movements, but they cannot eliminate them.

POLICY EFFORTS

This brings us to policy. What policies and programs should the federal government put in place to help family farmers survive in this uncertain, unstable industry? I will suggest three specific policy efforts. But in doing so, I recognize that the United States almost certainly would be unwilling to bear the heavy monetary costs, as well as the administrative misery, of trying to stabilize world prices of the basic commodities in the global market at levels acceptable to American farmers through such efforts as supply management, reserve stock management, or demand enhancement. My policy suggestions for implementation at the federal level are thus designed to help family farmers live and produce in an unstable industry.

Policy 1

Mount and support a strong extension program to educate farmers about the price and income instability of their industry and to advise them to pursue conservative investment strategies. Don't, for example, rush out and buy a new big tractor the first year commodity prices take a big upward jump. Remember that bad years always follow the good years. We just don't know when!

Policy 2

Develop a reserve program for the grains and oil seeds large enough to enable the U.S. government to (1) help moderate world price level movements by acquiring stocks in periods of falling prices and releasing them when prices rise; (2) serve as a reserve for combating famine conditions as they develop in various parts of the world (this could be most important as weather patterns change and conditions become more violent with global warming); (3) provide a secondary source of income to family farmers who provide on-farm storage for the government-owned stocks of grain.

Critics of the grain reserve concept will recall at once how the stocking operations of the Federal Farm Board in the early 1930s ended in failure. They may not recall, if they ever knew, that the large stocks held by the Commodity Credit Corporation in 1965–66 were used to advantage to ward off famine in India when crops were short there. But in truth, neither of the examples noted above, or numerous others that could be cited from the American experience in the past century, were legitimate grain reserve operations. The large stocks were acquired in open-ended operations to hold domestic grain prices steady or raise them above world prices. The stocks involved were accumulated in pure price-supporting operations.

In policy 2 I am proposing a legitimate grain reserve program in which stocks are acquired or dispensed within a rationally determined stock holding range. The magnitude of this range would be determined by the anticipated demands to be made on the grain reserve and the operating budget provided.

Of course, any open-ended commitment to support domestic grain prices above world levels is doomed to fail. But a legitimate grain reserve, as conceptualized above, with the goals outlined in policy 2, could play an important role in the uncertain world that lies ahead in the twenty-first century.

Policy 3

Develop a national sustainable agricultural policy and a set of programs, targeted at family farmers, with three principal objectives: give farmers an opportunity to earn additional income; help farmers develop operations that can more effectively withstand periods of price and income adversity; and improve the physical environment of rural areas for the benefit of farmers themselves and also the rest of society.

Two programmatic features are essential to this sustainable approach. First, for the farmers electing to participate in the program, each farmer working with experts provided under the program would develop a long-term sustainable plan for his or her operation; second, "green payments" would be made to each participating farm family for work undertaken under the plan and to cover expenditures made in the execution of it. As visualized here, a sustainable farm plan might call for the adoption of practices or operating procedures of the following kind: a reduction in the application of commercial fertilizers on a row crop; the increased production of a hay crop on the farm to be complemented by a beef cattle enterprise; the substitution of grassland or woodland for row crops at or near a water spillway, a stream or river, or a pond or lake.

This sustained policy approach would not be cheap. But neither have any of the federal farm programs since 1996 been cheap. This sustainable approach does, however, have something to show for the expenditures made—a restructured farming community that is environmentally benign, less emphasis on grain production for a phantom export market, and a farm economy that is more resilient—more capable of withstanding the adverse effects of severe price and income downturns that are certain to be a part of the unknown future.

For those who seek to understand what sustainable farming is all about and for those who have their doubts about the economic viability of sustainable farming, the case studies of three successful sustainable farms in the appendix should prove instructive.

One final comment is in order regarding the price and income instability of the farming industry. Should the commodity price decline of 1998–99 extend over a long period, say twenty years, as did the price decline in 1920–21, then one thing is certain: prices of farmland in the United States will fall—and fall to levels where U.S. producers are competitive with Latin American producers. Should that happen, chaos would reign in rural America, and the federal government would be forced to embark on bailout and resettlement programs beyond the scope of the policy proposals outlined above. This is not a prediction. No one can predict with certainty what will happen to the farming industry over the next twenty years. Perhaps changed weather patterns will so dislocate agricultural production around the world that supplies of the basic commodities will decline significantly and their prices will skyrocket. Once again, farming is a high-risk, inherently unstable industry, and both farmers and the nation had better be prepared for the unexpected.

THE TECHNOLOGICAL-BUSINESS ORGANIZATION PROBLEM

Let us turn now to the impact on family farmers of technological developments and changes in business practices. In the 1940s and 1950s when innovative farmers began adopting hybrid seed corn, their yields and profits increased. This led them to want to buy out their neighbors and expand their operations, which they did in increasing numbers. This farm expansion process was promoted by the adoption of large tractor hookups that enabled the ambitious farmer to cultivate more acres—and in fact required that he do so. Then a profitable crop, soybeans, came along, and again the successful, aggressive farmer wanted to expand his operation and alternate corn with soybeans. But with a fixed total supply of land, where could he acquire more acres? From his less successful neighbors, of course. The overall result for the period 1950–90 was an important increase in the average size of farms, an important decline in the number of family farms, and an important upward trend in the value of farmland.

This process has changed somewhat, perhaps slowed down, in the 1990s as the nature of technological advances in farming has changed. With the advent of computer technology in the information age, farm management has come to emphasize more effective employment of inputs in both farming and livestock operations. And new biotechnologies are changing the way farming and livestock operations are conducted. Computer technology in combination with new biotechnologies has made possible the efficient production of poultry, hogs, and milk in huge factories under controlled conditions. There are no family farms in these operations.

Where poultry and hogs continue to be produced in family-sized units, it is often under a contract that specifies how the birds or hogs are to be raised and how and where they are to be marketed for processing. As champions of market contracts often say, contracts take the risks out of animal and poultry production for the family contractor. But they can also reduce family contractors to the status of wage earners. The caretaker of a Tyson poultry operation might be likened to the manager of a fast food franchise. It's a job, and it may be a good job if the person involved has other good job opportunities. But it will be a poor job if his or her alternatives are poor.

On the farming side, new biotech inputs such as seeds are typically produced under patent by private firms. And through acquisitions and mergers these firms have been reduced to one or two giants. A farmer who wants to plant one of these patented seeds must sign a contract specifying how the seeds will be handled and perhaps how the crop will be marketed. Imagine the bargaining power of a medium-sized family farmer in southwestern Minnesota negotiating with a giant seed supplier like Monsanto. The farmer will have a choice: take the contract offered or leave it.

As of 2002, we have a farm structure in the United States that looks something like this: a few megaproducers of livestock and basic commodities that are large enough to either process their own products or bargain successfully with the buyers of their product; a few large processors who market their finished products under their own brand names and acquire their raw products (e.g., hogs, poultry) from many small units, each locked into producing them exactly as specified in the contract (the Tyson poultry syndrome); and finally the 500,000 or so "independent" family farmers described earlier who increasingly must sign contracts with an input supplier, a processor-handler, or both to survive. Our poor independent family farmer is being squeezed between giant input suppliers on one side and giant processor-handlers on the other.

It does not take much imagination to visualize an integrated food system starting with the retailer and running back through the processor-handler to the producer of the raw products (the farmer) and then to the supplier of the nonfarm-produced inputs. We are moving in that direction. The questions that remain are, Who will do the integrating, and how rapidly will it occur? And when it is complete, will there be one big monopolist, or can we look forward to at least some competition in the form of an oligopoly?

POLICY PROPOSALS

But before we lie down before this integration steamroller, what policies can and should be put in place to breathe life into the existing 500,000 or so family farmers? Three federal policy efforts are needed.

Policy 1

Public research and development in agricultural production, food processing, and human nutrition needs to be greatly strengthened so that an important share of new and improved technologies are in the public domain, not developed and sold by private firms under the protection of patent laws. The whole concept of public service to farmers and small businesses through education, research, and extension needs to be revived and supported with increased funding.

Policy 2

A special unit should be created in the Department of Justice to foster competition in the food and agriculture industries by rejecting mergers that lead to monopolistic positions and by prosecuting firms that engage in monopolistic practices (e.g., price fixing, issuing unfair or untrue information). This new unit should have special expertise in the new world of biotechnology, contract farming, and firm behavior under conditions of monopolistic competition and oligopoly; and it must recognize that bigness alone is an important aspect of monopolistic action.

Policy 3

The federal government should make an annual cash subsidy of $20,000 to each family farm grossing between $50,000 and $300,000 from the farming operation based on the family's income tax return.[2] This subsidy could be phased in below the minimum and phased out above the maximum in some reasonable schedule. It should not be tied to any particular commodity or quantity produced. It is being made to help these small to medium-sized commercial units stay in business when they must deal with large businesses on all sides, when they find it hard to obtain short-term credit and long-term capital on the same favorable terms as big business, and when they are at a disadvantage in most purchasing and selling transactions where product volumes and business connections play an important role. This is a subsidy to help these family farms survive when they must compete against bigness in every direction.

It is my thesis that family farms can survive with help. They may or may not produce as efficiently as some much larger farms. But their problem, on average, is not a lack of production efficiency. Their overwhelming problem is their disadvantage in daily commercial dealings with large-scale firms that have great market power, in which they—small to medium-sized family farmers with no market power—consistently come out second best.

The three policy proposals outlined above will help family farmers survive in the world of oligopoly and monopolistic competition in which they find themselves in this twenty-first century.

Some Concluding Thoughts

We have looked at two problem areas that have led to an untold number of business failures among family farmers over the years. And those problems continue to wreak havoc in 2002. To deal with those problems I have proposed a package of federal programs that in part are tested with experience and in part break new policy ground.

My proposed policy package does not come cheap, but neither do the farm relief programs of the late 1900s and early 2000s, which were put in place by Congress largely in the name of helping the family farmers, but whose funds go primarily to very large producers.

Family farmers can survive if they get the right kind of help from the government. But they are not getting it now and have not for a long, long time. The traditional commodity programs of price support and acreage controls based payments on the number of units produced; this has meant that most of the support went to the very large producers, who in turn could and did use their strengthened financial positions to buy out their less well-off neighbors. The phantom export market solution under the Federal Agricultural Improvement and Reform (FAIR) act of 1996 has resulted in a farm price and income crisis, well documented since 1997. The relief payments made to farmers since 1997 to compensate for the failed export policy solution, as noted above, have gone primarily to the very large producers.

It is time, I argue, to tailor the federal farm assistance package to the characteristics and needs of family farmers. This I have done in the proposals outlined above. I believe that the institution of family farming could survive and prosper under the farm policy package I present here.

AGRICULTURAL ABUNDANCE

Curse or Opportunity?

The downturn of farm product prices in 1997–98, and the continuing farm depression in the early 2000s, has the look of the 1953–66 downturn and depression. It was initiated by a contraction in the export demand, and it continues because of the continued pressure of abundant supplies on a restricted commercial demand. The future of commercial agriculture in America is thus far from clear. But we do know two things: the export market for the basic commodities is highly unpredictable; and conventional farming practices are, not so slowly, degrading the physical farm plant and the environment beyond the farm gate as well. Here I offer two visionary proposals for dealing with these continuing problems. My proposals may be visionary, but they are not new. The conceptual basis of both has been around for some time. I have simply given them new forms and, I hope, breathed new life into them.[1]

Some Historical Background

Farmers, farm leaders, politicians, and agricultural economists (including this one) spent much of their time in the twentieth century dealing with the curse of agricultural abundance—commodity surpluses and low farm prices. Policy proposals for dealing with this abundance problem were as varied as the individuals who made them. A return to the free market always had its advocates; but outside that group the dominant theme running through these many policy solutions was reducing the amount of farm product coming on the market. Reduce the supply, given the demand, and prices would rise, it was reasoned. The programs for achieving this reduction in supply sometimes involved restrictions or quotas on marketing, but most often they entailed some form of production control at the farm level. And for some reason, possibly because the participants realized this form of production control was a leaky vessel, the method of control most often adopted was taking land out of production.

The mechanics of these land-based control measures ranged from "skip-row" planting of cotton, which was totally ineffective in controlling production but much loved by farmers, to taking whole farms out of production, which was totally effective in controlling production but thoroughly hated by rural communities (it destroyed their population base). In the more usual case, a farmer received an acreage planting allotment for some crop, say wheat, that represented some percentage of his or her historical average acreage for that crop. The land not included in the allotment was to be held idle or planted to some cover crop. Sometimes farmers were paid outright to take land out of production, but more often the farmer was required to take a percentage of his cropland out of production to be eligible for price support or deficiency payments on the crop on his remaining land.

When commodity prices were low, which was often the case outside of wartime, farmers generally supported the idea of some form of production control. But in practice each farmer in the operation of his or her farm did everything that he or she could do to increase production and as a consequence sabotage the control program. The continuous introduction of new and improved technologies, together with the use of more intensive production practices on the allotted land, enabled farmers in many, if not most, cases to increase output on their individual farms. Thus the programs enacted to eliminate the surpluses and push up prices in the marketplace for the most part were not successful in controlling production, and hence in raising farm prices. The curse of abundance was thus never resolved satisfactorily in the twentieth century by the imposition of production controls. Mitigated perhaps, but not resolved.

In the early days of the New Deal the government acquired stocks of perishable commodities (e.g., dried eggs, cheese, potatoes) in various price supporting operations. It could not put these products back on the market, because to do so would push down the very prices it was trying to raise. And it could not destroy them to save storage costs, with millions of hungry people scrounging for food. So the government decided to do the obvious—to take this *opportunity* to give these surplus products to hungry people. Thus began a program of domestic surplus disposal that lasted several decades of giving whatever products the government had acquired in supporting farm prices to its own hungry people.

In the 1930s, a period of too much food supply and too little economic demand, there began to take shape in the fertile brain of one F. V. Waugh of the USDA a plan for increasing the economic demand for food products. He would issue food stamps to poor people that they could use to buy increased amounts of food at their local grocery stores; the government would reimburse the private stores for the value of the food stamps. An energetic doer in the USDA, one Milo Perkins, put the plan into operation in a few cities in the late 1930s, but with the onset of World War II those local operations were phased out. But agricultural abundance had for the first time created the opportunity to establish a program to feed people based on their nutritional needs.[2]

A companion program, the school lunch program, was also developed and placed in operation in the 1930s. The government first assisted schools by giving them surplus food products, but it quickly recognized that this was not a nutritionally effective way to support a national school lunch program. As the next logical step the government turned to providing financial support to schools to enable them first to get a lunch program started and then to subsidize free lunches for students too poor to pay. This program was continued through World War II and became an accepted national program. It has also spun off into several related programs: a school breakfast program, special milk programs, child care food programs, and a summer food service program. A special supplemental food program for women, infants, and children (WIC) was established in 1972 and has since grown into a major program serving over 4 million persons monthly. The 2001 budget funds the five children's food programs at a level of $10.3 billion and the WIC program at $4.2 billion. These child nutrition programs have become an important part of the USDA budget.

The food stamp plan was revived in the 1960s in the Kennedy-Johnson administration in a simplified form, but still on a limited basis. It was greatly expanded during the Nixon administrations and by the early 1990s had become the largest single program in the USDA, with an operating budget of nearly $15 billion in fiscal 1990. The food stamp plan continued to grow in the 1990s; the 2001 budget funds the food stamp plan at $21.2 billion, with participants estimated to reach 19 million.

Abundant food production throughout the 1900s created the opportunity to institutionalize food programs in the United States to protect and support children and to provide the poor with a regular supply of nutritious food. The curse of agricultural abundance was mitigated in part in the twentieth century by these domestic food programs.

With the ending of the enormous demands for food products created first by World War II itself and second, by the rehabilitation of war-torn countries, commodity surpluses and low farm prices re-emerged in the 1950s. Dramatic increases in crop yields and total farm output powered by new and improved technologies overwhelmed the weak production controls of the period, and stocks acquired by government in price support programs mounted to record-breaking levels. What were we to do with these surplus stocks that we could not consume at home or sell abroad? Perhaps stemming from our recent war and postwar experience of sharing our bountiful food supplies with our allies and defeated enemies, we hit on the idea of giving our surplus commodities away to poor developing countries around the world. Under programs officially identified as PL 480 and popularly known as "Food for Peace," for two decades or more the United States gave surplus commodities—primarily grains, but also some dairy products and vegetable oils valued at billions of dollars—to poor, friendly countries around the world.[3] In some cases we saved millions of people from starvation and death, as when short crops occurred in India in 1965–66, but in others cases we forced food commodities on poor nations as a way of helping them when they actually needed financial aid in their development efforts.

Our PL 480 programs were criticized by exporting competitors like Australia as one big dumping operation and criticized by conservative economists at home for reducing the incentive of these poor countries to increase their own agricultural production. Liberal economists, however, argued that these PL 480 shipments could and should be used as wage goods to further development programs in the recipient countries.[4] There is no question but that PL 480 was a surplus disposal operation on a huge scale; we disposed of what we had in surplus; a recipient country could take it or leave it. Most often poor countries took the commodities offered whether they fit their needs or not. There is no question either that our huge PL 480 operation, carried out over almost twenty years (1954–73) and at a cost to the federal treasury of $22.3 billion, enabled the United States to live with its agricultural surplus capacity without making any structural adjustments in its agricultural production plant. The PL 480 operation provided a way to live with the curse of agricultural abundance in America until something else came along—something else that would enable the United States to deal effectively with the curse of too much agricultural production.

That something was an increase in the commercial export of farm commodities, particularly grains, beginning in the 1960s but gaining strength throughout the 1970s and reaching a high in 1980–82. But commercial exports turned out to be a two-edged sword. The farm depression of the middle 1980s was driven by the decline in farm commodity exports in that decade, just as the farm prosperity of the middle 1990s was driven by an increase in exports in that period. The farm prosperity of the 1990s as well as the FAIR act were built on a tenuous export surge, which came to an end in 1997.

The Period 1997–2002 and Beyond

This has been an unhappy period for American farmers—especially those producing for the export market. Prices of the basic commodities (wheat, corn, and soybeans) fell sharply between 1997 and

1999 and remain low in 2002. As a direct result of the price decline in the basic commodities, cash receipts from crop farming are down, as of course are farm commodity exports measured in terms of value. But Congress has been generous to farmers in these troubled times; government payments to farmers from various assistance programs rose dramatically—to a high of $26 billion in 2000. As I write this in 2002, members of Congress are debating a new farm bill, and from all news reports they are prepared to keep voting huge subsidies to help farmers stay afloat financially—most of which go to big operators who are producing in large measure for the export market. To paraphrase one well-known advocate of global free markets, "We need to continue these subsidies until *something* comes along to strengthen those export markets." I hope that this *something* he had in mind is a strong recovery of the Japanese economy; in the twentieth century that *something* was most often a major war.

It seems that a sensible approach to dealing with the farm price and income problem would involve a reduction in aggregate production, and within that aggregate a reduction in the basics (wheat, corn, and soybeans), and as a consequence have less surplus products that must be exported. Along this line of thinking it is sometimes suggested that we return to paid acreage controls. If the acreage reductions are large enough and are tightly administered, aggregate output can be reduced. One advantage to this policy approach is that if export markets expand, the acreage controls can be dropped and the farm economy can return to full production in happy economic circumstances. There are at least two disadvantages: free market advocates among farmers, agribusiness firms, and economists hate production controls and will fight their imposition to the bitter end; and while the United States is reducing production by imposing acreage controls, producers in Latin America, South Africa, and Eastern Europe will be expanding their production and supplying the export markets once held by U.S. producers. Free market advocates are much happier living with cash subsidies and full production than with those hated production controls. But the question remains: How long will those cash subsidies keep flowing?

A second approach would involve a restructuring of the agricultural production plant in which farmers in highly productive areas like the Corn Belt would be induced through incentive payments—green payments—to farm less intensively and thereby reduce production, and cropland in marginal areas like the High Plains would be converted back to grass and used for some form of grazing. We have the opportunity now, in the early 2000s, to undertake this restructuring—to create a national agricultural plant where most farmers could, after a transition period, operate on a solid, sustained basis without a never-ending flow of subsidies. The funds are available to effect the transition (the funds now going to large commercial farmers as crop subsidies); our abundant production would cause no hardship to American food consumers; and American farmers would be freed in part from the ups and downs of an unpredictable export market.

Of course, this policy approach would be opposed by the same types who would oppose the paid acreage reduction approach, and perhaps by a few others. But I don't want to discuss its political chances at this point. That will come later, after a full discussion of the programs involved. Now I want to emphasize the opportunity that exists in the early 2000s to strike out in a new policy direction that focuses on protecting the physical environment and producing on a more sustained basis, and that as a consequence yields a bill of goods directed toward the food needs of American consumers rather than to an unpredictable export market. This will mean a smaller production plant, farmed less intensively than the present one, but one that is more stable and is sustainable over the long run.

THE SUSTAINABLE POLICY APPROACH

Sustainable farming is a concept whose time has come, but it remains poorly understood. John Ikerd provides us with a general but profound definition of a sustainable agriculture: "An agriculture that is capable of meeting the needs of the present while leaving equal or better opportunities for the future." He then elaborates on this concept:

> A sustainable agriculture must have three fundamental characteristics. It must be ecologically sound, economically viable, and socially responsible. Any system of farming that lacks any one of the three quite simply is not sustainable. This is not a matter for debate, it is just plain common sense. A sustainable agriculture must protect and maintain the productivity of its natural resource base. If the land won't produce, the farm is not sustainable. A sustainable agriculture must make sufficient profits to remain economically solvent. If the farmer goes broke, the farm is not sustainable. Finally, a sustainable agriculture must provide for the food and fiber needs of people, but it also must provide people with opportunities to lead successful lives. Agriculture must do its part to support society or society will not support agriculture.[5]

Here is a working definition of sustainability, consisting of three general principles that can be used to provide direction to a national program with the objective of moving as many farmers as possible to a state of agricultural sustainability:

1. *A sustainable farmer substitutes farm-generated or locally available production inputs to the extent possible for inputs produced outside the area (such as commercial fertilizers and chemical weed controls).*
2. *A sustainable farmer has adopted "positive" practices that diversify farm operations and provide alternatives to synthetic fertilizer and chemical use.* Positive practices include substituting animal and green manures for synthetic fertilizers, using tillage and crop rotation, and biological control of pests rather than chemical treatments.

3. *A sustainable farmer is committed to using locally produced inputs and to enterprise diversity.* Rather than perceiving sustainable practices as alternatives to use during difficult times, sustainable farmers are committed to changing their whole approach to agriculture, consistent with their new values.[6]

I propose here a national sustainable farm program to essentially replace the provisions of the FAIR act. Its purpose is to move farmers to a way of farming that has the capacity to provide them and their families with an acceptable standard of living, to protect and improve the quality of resources on participating farms, to put an end to the degradation of the environment beyond the farm gate, and to end the greater farm community's dependence on continuous farm subsidies. Conventional farm practices, and past and present government price and income support programs, have left the farm economy in a shambles and the land in a continuing state of environmental degradation. We need a new approach. I am proposing one: a national sustainable farm program.

The sustainable farm program, as I visualize it, would take the following form. The program would operate in areas where intensive cultivation practices are poisoning the land and water as well as causing soil erosion. These areas would certainly include the Corn Belt, the Central Valley of California, the Mississippi Delta, and many smaller high-production irrigated areas. It would be a voluntary program, but there would be important incentives to participate, both positive and negative. "Green payments" would be made to induce participants to move toward a sustainable plan of farm operation. On the negative side, transition payments and loan deficiency payments under programs in existence in 2001 would be eliminated. There would be both pull and push incentives to participate.

If some farmers elected to remain out of the sustainability program and produce, unsubsidized, for the free global market, they would be free to do so. They would, of course, be subject to regulations promulgated in conjunction with the sustainability program governing the management of animal wastes, the use of toxic materials, and soil erosion.

A participating farmer would develop, with the assistance of experts from the program, long-range plans for his or her farm designed to convert the farm to a sustainable operation.[7] A sustainable operation is one that ends the degradation of physical resources on the farm as well as beyond the farm gate, begins a steady improvement in the quality of resources employed, and provides the farm family with a living wage. Green payments would be made to the farmer for undertaking the actions required to bring the farm up to a sustainable operation. These payments would cover work undertaken under the plan, expenditures made in the execution of the plan, and any losses incurred in moving to a sustainable basis. These green payments should be viewed not as a permanent subsidy but as transition payments to help the farmer move his or her farm to a sustainable state. By definition a sustainable farm operation should not be dependent on continued government subsidies. The whole purpose of this exercise is to end up with farms that are sustainable in terms of both the physical environment and the economic environment.

To complement the action phase of the national sustainable farm program, there must be a second phase—a research and extension phase. Research is needed to develop new technologies and production practices that are consistent with sustainable farming systems and that can make sustainable farm operations more profitable. This is an area of research that has been largely ignored by the USDA and the colleges of agriculture over the years. This must be corrected by mounting a major research and development effort in support of a sustainable agriculture.

Many, if not most, American farmers do not have the foggiest notion about what is involved in sustainable farming. Thus, at the outset there must be an effective educational effort aimed at informing farmers of the objectives and mechanics of the program and, more generally, teaching them what sustainable agriculture is all about. I visualize an education program here comparable to what took place in 1933–34 when the first AAA programs were introduced to farmers. There will need to be meeting after meeting explaining to potential participants the nature of the program, the objectives of the program, what practices will qualify for green payments, and how to develop a farm plan. The extension service and the experts from the national program will first need to gear up for all the questions that will come to them and then take to the field providing answers to all those questions. It will be a demanding educational effort.

But the question may be fairly asked, Can these farmers practicing a sustainable mode of farming succeed financially when they no longer depend on the global commodity markets for wheat, corn, and soybeans as their principal source of income and are now concentrating on producing high-quality animal products and possibly specialty and organically grown foods? The answer is probably not, unless they can form cooperative marketing associations to seek out and find specifically defined or niche markets for their differentiated products—differentiated by quality and brands. This will take some doing. Thus, a third phase of the national sustainable farm program must take over at this point by assisting sustainable producers form marketing cooperatives to locate and supply these niche markets.

This assistance should take two forms: bringing enough producers together to form effective cooperative marketing associations, and helping the new cooperatives set up business organizations that can compete in the modern business world of bigness. The important point here is to create processing and marketing organizations that are owned and controlled by the sustainable producers but are managed by businessmen who know how to deal with the great retail outlets—grocery chains, restaurant and fast food chains, hotel suppliers, and the military. I do not recommend that the national sustainable program provide the capital to bring these cooperatives into existence. *They should be capitalized by their producer-members.* The financial success of such cooperatives will depend in large measure on the patronage support of their producer-members, and that support will be strong when the producer-members have a financial stake in the success of their association. But the initial organizing phase for these cooperative marketing associations will be difficult, and that is where the National Program can be helpful in providing subsidized operating capital, business organizational know-how, and informative accounting procedures.

The operating objective of these cooperative marketing associations should not be to gain monopoly positions in some part of the marketing channel. It should be twofold: to locate and supply the niche markets for the high-quality products of its sustainable producer members, and to be big enough and strong enough to withstand the competitive hits it will take while operating in the food marketing system.

The recent experience of hog producers in the Corn Belt in coming together in cooperatives to slaughter their own hogs and market the pork products has not been good. The specialized, or niche, markets are there, but the retail operators in those markets are hesitant to break away from established suppliers and take a chance with the products of new farmers' cooperative suppliers. Thus, there is an important, and absolutely necessary, role for the national sustainable farm program to play in assisting these fledgling processing and marketing farmers' cooperatives in their first years of operation.

In 1991 the Northwest Area Foundation supported a study of the economic performance of conventional farms and sustainable farms in four states (Iowa, Minnesota, North Dakota, and Montana).[8] Financial and performance measures for conventional, sustainable, and state average farms are presented in considerable detail in the report. The year 1991 was a poor year financially for all farmers; on average none of the farmers in the survey, conventional or sustainable, earned enough to pay for family labor and pay themselves a market rate of return for the net worth they supplied to the farm. The survey data show that sustainable farmers did not perform as well financially as their conventional neighbors in that year. But because of the diversity of farms and farming styles in each state, there was

more variation within conventional and sustainable groups than between them. For example, in three of the four states, the top one-third
of sustainable farmers earned respectable returns on net worth. (See
the tabulation below.)

	Iowa	Minnesota	Montana	North Dakota
Number of farms	18	7	9	13
Return on farm net worth (%)	7	0.6	5	11
Net farm income	29,997	19,376	70,435	49,937

Many of the Minnesota farms in the tabulation are dairy operations, a sector that generally fared poorly in 1991. But even Minnesota,
which comes off poorly in 1991 in terms of returns on net worth, had
some financially successful sustainable farms in 1992. (Case studies of
three successful sustainable farms in Minnesota in 1992 are presented
in the appendix.)

The evidence on the economic performance of sustainable farms
from the Northwest Area Foundation study is mixed: some did well,
more did poorly. The wide variability in their success suggests that
sustainable farming may present a greater management challenge
than conventional farming, and the need for information and technical assistance support among sustainable farmers is high. This is
recognized in phase two of the proposed national sustainable farm
program.

Having recognized that achieving widespread financial success
among participants in the national sustainable farm program will require a major educational effort, let us further recognize that these

farms, as well as those outside the program, will be subjected from time to time to the kind of downswings in the prices of basic commodities that occurred in 1998–99. However, farmers who have achieved a high state of sustainability will be in a stronger position than their conventional neighbors to withstand sharp downturns in product prices, since they will have reduced their dependence on nonfarm-produced inputs and will, in many cases, be selling their product to niche markets, both domestic and foreign. The move toward sustainability in farming represents an important step toward making an inherently unstable industry more stable.

To help all farmers deal with severe price fluctuations, I argue once again for two things. First, we need a grain and oil seed reserve to help moderate these great price swings in the basics, as well as to provide farmers with an added source of income in the form of storage payments. Second, a federal refinancing agency should be established to help family farmers refinance loans at subsidized interest rates or relaxed payment schedules in periods of falling commodity prices. (This is critical to young farmers just getting started and carrying large mortgages to finance the purchase of land and other large items.)

In chapter 6 I argued for an annual cash subsidy of $20,000 for each family farmer grossing between $50,000 and $300,000. The national sustainable farm program becomes a substitute—and I believe an effective substitute—for such a cash subsidy program for small to medium-sized family farmers, just as it takes the place of the payment programs under the FAIR act. If, however, the national sustainable farm program never becomes a reality and the conventional payment programs based on the volume of production of each farmer are continued, then I once again argue for an annual cash subsidy to small to medium-sized family farmers. To survive in the business world of bigness, they need help. And this cash subsidy of $20,000, not tied to any particular commodity or farm practice, provides that help.

The policy approach being proposed here is straightforward: bring to an end as quickly as possible intensive cropping on marginal lands and return those lands to their natural state. On the High Plains this means returning those lands to grassland, seeded with native grass species where possible, then using them for grazing. Marginal cropland in the mid-South and Deep South might, for example, be returned to timber. I suggest this policy approach for dealing with marginal cropland throughout the United States; but in this essay I will focus on the High Plains, since I know that area best.

Chouteau County, located in central Montana astride the Missouri River, was once the bison hunting ground of the Blackfeet Indians. Then came the fur traders, followed by the steamboats up the Missouri loaded with trade goods; next came the gold seekers, then the military to subdue the warring Blackfeet, followed by the cattlemen bringing their herds of cattle up from Texas to exploit the wonderful High Plains grass and almost to face extinction in the blizzards of 1886. Then the wheat farmers came to homestead the land in the early 1900s, reap a bonanza during World War I, go broke in the 1920s, and starve out in the 1930s. Those who were left were able to enjoy wartime prosperity in the 1940s and early 1950s, experience some good times and bad times in the 1970s and 1980s, and now finally, at the dawn of the twenty-first century, come down to living off the largesse of the federal government.

Timothy Egan sums up the situation in two short paragraphs:

> What has happened in rural counties like Chouteau completes a full circle, from the creation of farms by government incentive through the Enlarged Homestead Act of 1909 to a period of prosperity and independence in the 1950s and 60s, to the present where government is the only thing keeping people on the old bison grounds of half of Montana.
>
> The homesteads have become sources of export crops. Nearly 90 percent of the wheat grown in Montana is sent

overseas. But it faces global competition and a glut. Even countries like Pakistan, once seen as a relief target, are now exporting grain. If the Montana growers were to try and get by in the free market, they would lose about $2 on every bushel of wheat they grow.[9]

Grain farmers on the High Plains from the Canadian border to the Rio Grande are in a financial fix similar to the one just described for Chouteau County, except where they have access to cheap water. With water farmers can diversify and grow anything that shows some chance of yielding a profit—perhaps cotton in the South, maybe sugar beets or alfalfa in the North. But cheap water in the southern plains based on pumping the Ogallala aquifer is not so slowly coming to an end as water levels in the aquifer keep falling and the costs of pumping keep rising, especially at the southern end of the aquifer. As of 2002, the end of that water bonanza is in sight.

What then to do with these dryland wheat farmers and the small towns, machinery dealers, school systems, and churches they support? The politicians who represent the High Plains people say we should continue to pay out the cash crop subsidies. The politicians who represent the rest of us either lack the courage to turn off that subsidy spigot or lack the vision to correct the mistakes that began a century ago and continue today.

The Poppers had it right back in 1987 when they said that large parts of the Great Plains should be converted into a fenceless "Buffalo Commons."[10] I propose that this land, outside sustainable irrigation oases, be converted back to grass and then organized into large units of three possible types: cattle ranches, bison or buffalo ranches, and ecological reserves for wildlife. The type of unit selected for a specific area will depend on numerous conditions and criteria: geographical location, contiguity of the parcels of land, national environmental priorities, historical commitments, and on and on. But before trying to sort out these management questions, let us turn to the possible content of a program to effect this restructuring.

1. The geographical area in which the restructuring is to take place must be defined, and the basic objective of the program must be made clear to all concerned. (This is a critical first step and may kill the program before it even gets under way, but unless the federal government can take this step there can be no effective program.)

2. The farmers, as well as all the other citizens of the area, should be informed that all cash crop subsidies to farmers are to be terminated within one year after the announcement of the program. This is to provide the push to get the restructuring moving.

3. A voluntary farm buyout program should be placed in operation with instructions to pay farmers for their land at levels that existed one year before the announcement of this program. There is obviously some subsidy in such a purchase price, but it is put in the proposal first to provide a pull to get the restructuring moving and second to help dislocated farm families get a new start in life. Those farmers not electing to sell would be free to continue farming and to operate in a completely free market situation.

4. In conjunction with the farm buyout operation, a relocation operation should be put in place. This program should help both farmers and nonfarm people to relocate outside the program area by providing information about job opportunities outside the area; granting scholarships to enable dislocated people to learn new work skills; and making loans to help dislocated families relocate in new areas.

5. After, say, two years of farm acquisitions, the program managers would need to inventory these holdings, take stock of those lands (their geographic location, their contiguity, their total acreage), and begin to establish procedures for redistributing the acquired land into some kind of utilization units. Those procedures would, of course, be governed by priorities and criteria set forth in the enabling

legislation creating the program. Those priorities and criteria are all-important and must be clear and concise if this land redistribution process is to avoid the multitude of corrupt practices associated with the original distribution of the public domain. In my judgment the following criteria should govern the redistribution: (a) A significant amount of the acquired land (the exact amount to be set forth in the enabling legislation) should be *given* to the Native American tribe or tribes adjacent to the land, to be held by them in perpetuity and to be managed by the governing body of the tribe for use in raising and grazing bison or other livestock. (b) Where the acquired lands can be aggregated into a unit of at least 3,000 square miles (perhaps in combination with existing federal lands), ecological reserves should be created on the High Plains capable of supporting on a sustained basis thousands of bison, elk, mule deer, antelope, several wolf packs, some mountain lions, and bighorn sheep (terrain permitting), and a million or more prairie dogs as well as their natural predators, the black-footed ferrets.[11] Perhaps even some grizzly bears, which once roamed these plains, might be included. There is no reason we cannot have our own Serengeti Plains here in America. We have the animals, and we could have the space; all we lack is the will. (c) Parcels of acquired land that do not fit into the categories above should be rented at prevailing market rates to adjacent cattle ranches for grazing. After experience is gained with this restructuring program, consideration should be given to selling the rented lands to the cattle ranchers, with the proviso that they must be kept in grass and used for grazing.

6. Questions concerning who should administer this program and where the agency doing the administration fits into the federal scheme of things are knotty ones. In the early phases of the program questions concerning the area

boundaries and farm purchase agreements are largely agricultural; in later phases questions raised about tribal governments and ecological reserves fall largely in the domain of the Department of the Interior. And from the beginning, high-level politics will play a major role in creating the program and setting the priorities of land redistribution. Thus I propose that the program be administered by an independent agency with the chief administrator reporting to the president's chief of staff. After the agency has been operating, say, ten years and has met, we hope, with considerable success, it might be brought under the wing of the secretary of the interior, and at some later date its various functions might be distributed among the concerned bureaus of the Interior Department: Indian Affairs, Land Management, and Fish and Wildlife. But that is a long way down the road. The first and greatest problem will be bringing this restructuring program for the High Plains into being.

The course of action outlined above may seem radical, and perhaps it is. But the key ideas are not new. In 1878 John Wesley Powell wrote this in his famous report to the Congress, *Report on the Lands of the Arid Region of the United States*:

> The farm unit should not be less than 2,560 acres; the pasturage farms need small bodies of irrigable land; the division of these lands should be controlled by topographic features to give water fronts; residences of pasturage lands should be grouped; the pasturage farms cannot be fenced—they must be occupied in common.
>
> The homestead and pre-emption methods are inadequate to meet these conditions. A general law should be enacted to provide for the organization of pasturage districts, in which the residents should have the right to make their own regula-

tions for the division of the lands, the use of the water for irrigation and for watering the stock, and for the pasturage of lands in common or in severalty.

He was trying at the height of the homestead era to head off the land use pattern on the High Plains that we are now taking steps to correct. His vision for the High Plains was that of one great fenceless pasture, dotted here and there along streambeds with ranch headquarters. But he was unsuccessful; Congress could not bring itself to give a homestead of 2,560 acres. The mental and legal views of the lawyers and politicians based on their experience in the humid East (east of the ninety-eighth meridian) were too strong to be overcome.[12] So, well over a hundred years later we must entertain radical ideas once again to deal with the difficult economic problems of the arid High Plains.

The Realities

I have argued that in 2002 and beyond the opportunity exists to restructure the American farm sector, first along sustainable lines in the productive Corn Belt and second by turning marginal wheat land on the High Plains into grassland. My critics will say that my visions of the future ignore political realities. My response is that my visions are based on economic and environmental realities. Whether through attrition or through conscious collective action, those economic and environmental realities will determine the ultimate productive structures of the Corn Belt and the High Plains.

Through attrition, the High Plains have been moving from crop farming to grass and grazing over the past fifty years or longer. This process of attrition is real, but not pretty. Let Florence Williams describe it for you:

> In Souris, N.D., just shy of the Canada border, no one is in at the Cenex gas station. A bulletin board advertises a "canola risk management meeting" and haircuts for $7. The post office, hardware store, grocery, and café all share the same

low-slung aluminum-sided, red-roofed building. No one is there, either. This is Bottineau County, which had 17,295 people in 1910. It had 7,241 people, and dropping, in 2000. In the 1990s alone, 10 percent of the county either died or moved. Two counties to the West, Burke County lost 24.5 percent of its population in just a decade. . . .

It's not that this land is empty; it's that it's post-industrial. Weeds have taken over the road-sides, piles of rusting sheet metal sit in fallow fields, and even the oil derricks have squeaked to a halt. This is the South Bronx of the prairie. To be redeemed, it also needs to be restored.[13]

Two conditions have kept this process of attrition from turning into an avalanche of change: the top-heavy political representation in Congress from the plains states and the pumping of the Ogallala aquifer on the southern plains. The Ogallala aquifer is being drawn down to levels where it is too costly to pump that water for irrigation. And certainly at some point the nation will become aware politically that it makes no sense to subsidize the production of high-cost wheat on the High Plains when the market for it either is nonexistent or has become completely unreliable. When these conditions are realized the High Plains, apart from some limited sustainable irrigation districts, will return to grass and grazing, as nature intended.

The Corn Belt is a different story. It is a highly productive area, producing great volumes of corn and soybeans. It is to the advantage of suppliers of nonfarm produced inputs (chemical fertilizers, herbicides, and biomodified seeds) that farmers continue to produce in great volume and hence continue to buy these inputs in large quantities. It is also to the advantage of marketers (traders, speculators, processors) for farmers to continue to produce large volumes of corn and soybeans, on which they earn a return on each bushel handled, and sometimes a handsome return. These large, powerful agribusiness firms regularly join forces with the few remaining farm congressmen to lobby both Congress and the administration (regardless of party) to continue to subsidize the highly productive Corn Belt farmers. With the cash subsidies flowing, the farmers continue to pour on the nonfarm-produced inputs (largely chemicals), and the large product volumes keep moving to market. Up to now most people living and working in the Corn Belt have been satisfied if not happy with this state of affairs (farmers are never happy), and the input suppliers and marketers and their political allies have been overjoyed.

But there is a dark side to this picture, which is emerging around the edges. A good number of farmers are learning that their groundwater is being poisoned from too great a concentration of nitrate. The runoff from cropping areas where applications of nitrogen have been heavy is polluting and poisoning streams and rivers all the way to the Gulf of Mexico, where a dead zone develops annually in which shrimp, fish, and other marine life cannot live. Topsoil erosion from continuous cropping continues to be a serious problem in the Corn Belt.[14]

Gyles Randall, a soil scientist at the University of Minnesota, reports that corn and soybean acreage, which occupied 64 percent of the crop acreage in southeastern Minnesota in 1975, had increased to 80 percent in 1999. This shift to greater corn and soybean acreage has been accompanied by fewer and larger farm operations, fewer livestock farms, more pest problems, increased iron chlorosis, and increased soil erosion. These developments lead him to question "whether present day agriculture in southern Minnesota is sustainable from economic, environmental, ecological and sociological perspectives."[15]

In this not too happy context an increasing number of corn and soybean producers are asking, Why do we go to such lengths to increase production when the markets for our commodities are weak and unpredictable and we can see the environmental problems piling up? The answer, of course, is that they are told on every front (college extension agent, agribusiness representative, farm magazine) that they must get bigger and produce more to survive. That is the prevailing business culture, conventional business wisdom, in the commercial farm sector.

In this business climate agribusiness will continue to push farmers to increase production by providing new and costly technologies and inducing government to give them cash subsidies to produce more corn and beans. Increased volume is their business. Elsewhere I have likened a large, modern Corn Belt farm to a huge sausage machine—add ingredients at the front end, run the machine, and product flows out the other end. In the case of the Corn Belt farm, make generous applications of nonfarm-produced inputs (e.g., chemical fertilizers, specially designed seeds, tillage, planting and harvesting machines) to the land, operate the farm through the growing season, and at the end of the season harvest a bountiful crop of corn and soybeans. Depending on input costs and product prices, in some years this corn- and soybean-producing machine may earn a respectable return; but in more and more years it must be subsidized to keep it running.

This process will continue unabated until Corn Belt agriculture runs into a big crisis. That crisis may be environmental, or it may be economic in nature. The early stages of both are taking shape as I write this essay. When the crisis, or crises, becomes full blown, the pressure will develop to take action—to move in a new policy direction. At that time the ideas presented here just might take hold.

Once again our farming sector is in trouble, as it has been many times in the past century. It is not in trouble because it can't produce. It is in trouble because it produces too abundantly. The agribusiness solution to the problem is to produce still more. The congressional solution is to throw more money to the farmers—that is, to the big farmers. Conservative economists focus on minuscule marketing or production problems and hope that *something* big will come along to once again raise farm prices. I have suggested that there are some real environmental and economic problems involved in this latest period of farm troubles. And I have provided some visionary collective actions to deal with those problems. We will see what happens over the next decade.

EPILOGUE
The Future?

No one knows how long the current farm depression will last. The optimists—the free market globalists—see an expanding world economy, an expanding global demand for the basic farm commodities, and good times once again for American farmers just around the corner. The pessimists—old-time Keynesians like me—see slow-growing national economies, a weak global demand for the basic farm commodities, and hard times for American farmers stretching into the uncertain future. Some foreign observers, regardless of the world economic situation, see low-cost feed grain and soybean producers in South America whipping American Corn Belt farmers in head-to-head international competition, resulting in sharply falling farmland prices and widespread business failure among Corn Belt farmers.

Well-informed writers on global warming see weather patterns around the world changing, in some cases drastically, perhaps disrupting agricultural production to an important degree. There could be food shortages in broad areas of the world, with skyrocketing commodity prices in global markets, but there might also be physical and economic disaster for farmers over an entire region. In all this uncertainty, what will Congress be doing? Increasing financial aid to U.S. agriculture? Or whittling down that aid? And will that aid, such as it is, go primarily to the relatively few very large, well-established farmers, as it has in the past, or could Congress surprise us and start directing much of that aid toward the small to medium-sized family farmers? Who knows?

As matters stand today, the future for American farmers, particularly family farmers with modest financial reserves, appears to be one big crapshoot. Commercial farmers with high fixed costs in land and machinery face increased low-cost competition from abroad in an unreliable export market and are dependent for survival on cash subsidies from the federal government, year after year. There are, however, some certainties in the present and the future. The physical environment on farms, and in the surrounding areas in intensive farming regions such as the Corn Belt, is not so slowly being degraded and in places made destructive to plant and animal life. And the existing national farm production plant is so productive, and the surplus production capacity so pervasive, that the opportunity exists to restructure that production plant along sustainable environmental lines without in any way endangering the food supply of the national society.

How is this to be done? It is to be done along the lines described in the previous chapter: by converting the High Plains cropland back to grass and grazing operations and by transforming intensive cropping areas like the Corn Belt to diversified farming areas making greater use of livestock operations, making less use of nonfarm-produced inputs (e.g., chemical fertilizers and pesticides), and producing differentiated products for niche markets both domestically and abroad.

Doesn't the adoption of these production policies and practices represent a step backward, technologically, you ask? The answer is yes if by technology you have in mind just big machines and chemicals. But it represents a step forward if you have in mind the adoption of practices and inputs that enhance the physical environment, both on the farm and beyond the farm gate. The development and operation of a sustainable farming system does not mean that farmers must employ antiquated practices—that they can't, for example, adopt new and improved plant varieties. It simply means that, whatever farming practices are adopted, they are in harmony with the physical environment and sustain and improve it over time.

Where is the money coming from to transform the High Plains into a grazing economy and the highly intensive Corn Belt agriculture into a diversified, sustainable agriculture? From the same money that is now being used to subsidize the intensive farming systems year after year. This transformation will not come cheap. But it need not cost any more than the present farm subsidies that are running to $30 billion a year. But there is a difference. After five or ten years of such a government-supported transformation process, we would end up with a restructured farming system—a sustainable system. With the present system of the government subsidizing of intensive farming, no one has the foggiest idea of where, when, or how it will end. Vocal supporters simply hope that "something big" will come along to bring it to a happy ending.

Finally, you ask, do you really believe these diversified, sustainable farms that are to take the place of existing, intensified farming operations can stand on their own feet without a continued government subsidy? My first answer is, We'll never know till we give it a try. My second answer is positive, for several reasons. First, these sustainable farms will make less use of nonfarm-produced inputs, hence their annual operating costs will be much lower than for existing conventional farms. Second, with most farm production occurring on these sustainable farms, their aggregate output will be less than for a comparable set of conventional farms, hence the pressure of supplies on demand will be less, hence the level of farm prices should, on average, be higher. Third, these sustainable farmers will be producing for niche markets, tailoring their products in quality and volume to the needs of those niche markets, and so enjoy the financial advantages of operating in markets where they can adjust supplies to the demands of those markets.

For these reasons, I believe that sustainable farming in the United States can be economically viable. Fortunately, we have the opportunity *now* to move to an environmentally sustainable system—our production abundance provides that opportunity. If we take advantage of it, American family farmers can look to the future not with despair, but with realistic hopes of building successful farming operations for the long haul.

APPENDIX | WHAT MAKES SUSTAINABLE FARMS SUCCESSFUL?

Case Studies of Three Minnesota Farms

Jodi Dansingburg, Charlene Chan-Muehlbauer,
and Douglas Gunnink

Can environmentally sound farming contribute to financial success? For three south-central Minnesota farms studied during the 1992 crop year, the answer is yes. These farms use little or no chemical fertilizer or pesticide, employ other soil-building and environmentally beneficial strategies, minimize capital cost, and emphasize net return more than gross production. The result is an economic return superior to the conventional farms in their area.

We compared these farms with average farms in an area association, the South Central Minnesota Farm Business Management Association (SCMFBMA). Although gross income for two of the three was substantially less than average SCMFBMA farms, *net profit was much greater*. The high net profits were achieved because profit margins (percentage of gross farm income retained as profit) were greater than an average of the ninety highest-returning SCMFBMA farms.

Further, these sustainable farms achieved their success on less than half the acreage of the highest-returning 20 percent of SCMFBMA farms. Two of the farms also did it with fewer cows than average for the region.

Because each farm has unique resources and obstacles and each farmer has distinct talents, interests, and management ability, there is no single recipe for a successful farming system. Nevertheless, these farms demonstrate the potential for every farm to optimize productivity and biological efficiency and to improve profits.

Two themes run through these farming operations: *emulating natural systems* and *minimizing production cost*.

Profitability of Three Minnesota Sustainable Farms Compared with Area Averages for the Crop Year 1992

Farms	Gross cash farm income[a] ($)	Net farm income[b] ($)	Profit margin[c] (%)	Size of farm (acres)	Number of head[d]
Webster[e]	111,988	63,622	60.4	240	40
Mason[e]	263,508	43,423	21.0	248	75
Elwood[e]	96,683	43,275	48.4	287	28
South-central Minnesota average[f]	205,832	28,391	13.3	484	53.5
South-central Minnesota highest-returning 20%[g]	340,641	67,219	18.4	807	53.2

[a]Excludes off-farm income. [b]Inventory-adjusted profit. [c](net farm profit)/(value of farm production) x 100. [d]Average number of head on farms with a dairy enterprise. [e]A pseudonym. [f]Average of 448 farms reported in south-central Minnesota Farm Business Management Association annual report. [g]Average for the highest-returning 20% of the 448 participating farms reported in the south-central Minnesota Farm Business Management Association 1992 annual report.

Using Natural Systems as a Model

In nature, species interact in systems, the way bees and flowers work together to manufacture honey. All three farmers work *with* these systems, not against them, to assemble their crops and animals into profitable farms.

Keen Observers

In refining their operations to work better with natural systems, all three operators have sharpened their *observational* skill. Each described how he has based many management decisions on experimentation and observation of soil, weeds, crops, and livestock. Bob Elwood (all names used here are pseudonyms) described how he decided to feed minerals "free choice" to his dairy herd rather than force feeding: "I just kept backing the minerals out of their feed and watched how interested the cows were in eating. The more I backed off on the minerals, the more feed they ate. I can't afford not to take the time to observe these things [soils, crops, weeds, and livestock]. If I farm on a smaller scale [and take the time to observe], I'll make more money."

By experimenting with different soil, weed, crop, and livestock practices and carefully observing the results, each operator has developed integrated farming systems that create healthy environments and profits. Another operator, Kevin Webster, noted: "I don't care as much about NPK levels [levels of nitrogen-phosphorus-potassium chemical fertilizer] as I do about soil microbes and soil health. These organic fertilizers [animal and green manures] slowly release minerals into the soil. On warmer days, the soil microbes are more active, so more minerals become available in the soil. It so happens that on those same warm days, plants are growing faster and require more nutrients. See how Mother Nature fits things together?"

All three farmers noted that their soil structure has improved dramatically since they replaced chemicals with the judicious use of crop rotation and managed application of green and animal manures. Their actions have improved the soil's ability to hold water. Now their soils can absorb rainfall of over two inches without noticeable run-off, whereas neighboring fields, farmed conventionally, have "streams between every corn row." The third farmer, Jack Mason, observed: "There's something in [my] soil that allows it to hold the water that other people have destroyed by using some of their methods—by some of the [chemicals] they've put on."

Drought tolerance also appears to be improved by maintaining healthy soils. The farmers commented that their crops appeared to weather recent droughts better than those of their conventional neighbors.

Humane Treatment of Livestock

The three farms described here also have modified their livestock production practices to simulate natural systems. Mason and Webster both are adapting controlled grazing systems to their dairy herds. Webster also raises his pigs on pasture. The animals harvest their own feed, spread their own manure, and stay healthy through exercise and fresh air. "You have to follow the lead of the animals," said Webster. "If you watch them, they'll show you what housing and feeding systems they prefer."

Ecological and Agricultural Diversity

A key to using natural systems as a model is *diversity*. Natural eco-systems have diverse soils, plants, and animals that help maintain water and mineral cycles. Each of the three farmers raises varied crops and livestock, creating more natural ecosystems and giving them flexibility against variable market prices:

> Kevin Webster has forty dairy cows and a farrow-to-finish hog enterprise. He also markets organic soybeans and small grains.

Jack Mason has seventy-three dairy cows (registered Holsteins) and raises and sells all his registered bulls and unneeded heifers.

Bob Elwood has twenty-eight dairy cows. He sells off unneeded heifers and markets male calves as beef steers (half the steers are sold directly to consumers, thus increasing profit). He sells some corn and oats and markets organic soybeans and blue corn.

As Kevin Webster points out, it is the very diversity of his crops and livestock that allows him to piece together a farming system that is productive yet not overly demanding.

Crop Diversity

Most corn-soybean producers participate in government programs and receive subsidy payments for their corn. This payment is calculated on their *base acres*, the area historically planted in corn. Consequently these corn producers avoid rotating into other crops because they lose base acres, resulting in smaller payments.

All three farms are in a traditional corn-soybean rotation region. Yet each of these sustainable farmers has made his farming system more profitable by expanding crop rotations, such as growing a mixed crop of field peas and small grain (oats or barley) as part of the rotation. This breaks weed and disease cycles, provides ground cover for erosion control, and does so at a very low production cost. The peas add enough protein to the small grain that the harvested crops can be fed to their livestock with little or no supplemental protein.

Kevin Webster's livestock gain weight and produce milk so well on this feed that he believes he can raise as much beef, pork, or milk from an acre of barley and peas as from an acre of corn, despite corn's greater yield.

One way to encourage more farmers to use beneficial rotations is to change the commodity program to let them plant beneficial crops without penalizing them with loss of base acres.

Minimizing Production Cost

Successful limiting of production cost has two components:

1. Minimizing purchased inputs, such as fertilizers and herbicides.
2. Strategically minimizing capital investment.

Minimizing Purchased Inputs

All three farmers have reaped the benefit of emulating natural systems. None purchase commercial fertilizer. Two of them use no herbicide, and the third applies minimal herbicide when mechanical weed control is difficult. In addition to applying manure before planting corn, all three plow in an alfalfa crop for additional fertility.

By relying on farm-generated fertility and pest control, all three farmers saved over $50 an acre on fertilizer and chemical cost on their 1992 corn crops, compared with the average farm in their region.

Despite the cool, wet growing year in 1992, these three farmers had an average net return per acre on corn of $121, comparing very favorably to the area's average net return per acre on corn of $-\$7.47$ and the average of $35 for the top 199 producers. Two of the three farmers achieved their high net returns on corn *without* the assistance of government program payments.

Minimizing production cost carries through to these successful operators' livestock enterprises. All three spend less than the average south-central Minnesota farm on grain and forage for dairy cows, again emphasizing profit over production.

They also believe that restricting chemical use on crops keeps their livestock healthy. Two of the three had below average veterinary cost for 1992; one spent $31 in 1992 for the entire dairy herd. For these operators, the 1992 net return per cow averaged $850, compared with the SCMFBMA average of $387 per cow.

Strategically Minimizing Capital Investment

These farmers also limit investment in livestock facilities:

> Kevin Webster pasture-farrows his sows in the summer (they
> have their litters in the field rather than in pens or barns).
> He uses controlled grazing for both his hogs and his dairy
> animals.
>
> Jack Mason finds it more cost effective to hire a neighbor
> to harvest his corn than to invest capital in a combine—
> plus paying interest on the loan, and maintenance, and
> repairs.
>
> Jack also uses controlled grazing to feed his young stock in
> summer.
>
> All three farmers have minimized their investment in feed-
> storage equipment.
>
> All three have older field equipment to minimize overhead.
>
> All three have kept their debt levels low by limiting farm size. All
> the farmers bought their farms, and two of the three have
> paid off all their land debt by using practices described
> and without the benefit of organic premiums.

By minimizing production cost, these farmers have given themselves resiliency against fluctuating market prices. This resiliency allows them to make a profit under broader conditions than their
conventional counterparts.

The three farmers in this study are successful because they manage
their farms *as a sustainable whole, not as pieces.* Although each enterprise is environmentally and economically successful, *the overall
strength of their farms comes from the synergistic interaction of their
soil, crop, livestock, and financial management systems.*

By minimizing production cost, focusing on increasing profits
instead of increasing yield, and using natural systems as a model,

these three sustainable farms have attained net farm income well above the regional average. Their success challenges the conventional wisdom that net farm income can be increased only through increasing gross farm income by expanding acreage (requiring capital for land) and livestock confinement.

Source: Planting the Future, edited by Elizabeth Bird, Gordon Bultena, and John Gardner (Ames: Iowa State University Press, 1995).

1. The Case for Production Control

1. The annual meeting of the Metropolitan Cooperative Milk Producers Bargaining Agency, Syracuse NY, December 1954.

2. The Agricultural Treadmill

1. Willard W. Cochrane, *Farm Prices: Myth and Reality* (Minneapolis: University of Minnesota Press, 1958).

2. For a good discussion of the unequal rates of growth between aggregate demand and aggregate supply, see T. W. Schultz, *Agriculture in an Unstable Economy* (New York: McGraw-Hill, 1945), chap. 3.

3. This amount does not include funds used to finance research on the handling, processing, and distribution of products after they leave the farm.

4. James T. Bonnen, "Adjusting the Structure of Agriculture to Economic Growth," paper presented before the North Central Farm Management Research Committee, Chicago, March 18–20, 1957. The full report of this major study is to be published by the National Planning Association.

5. This section is adapted in part from W. W. Wilcox and Willard W. Cochrane, "Who Gets the Benefit of Farm Technological Advance," in *Economics of American Agriculture* (Englewood Cliffs NJ: Prentice-Hall, 1951).

6. Dale Kramer, *The Wild Jackasses* (New York: Hasting House, 1956).

7. Countless alternatives falling between these extremes could be considered.

8. It is sometimes argued that increases in total output between 1920 and 1955 may be attributed to farm firms' becoming more efficient over this long period (i.e., farmers were successful in locating and moving toward the minimum point of their long-run planning curves). But this argument contravenes facts and logic. The state of the art was changing over the entire period, sometimes slowly, sometimes rapidly. Farmers were constantly adjusting to new levels

and patterns of technology. By what logic, then, can one argue that farmers were more nearly at the minimum point of their long-run static cost curves in 1955 than farmers were in 1920? None, except by assertion. The facts are that farmers were adjusting to new technologies over the entire period, not seeking minimum points on *static* planning curves.

9. John M. Brewster, "Farm Technological Advance and Total Population Growth," *Journal of Farm Economics*, August 1945, p. 515.

10. J. R. Bellerby, *Agriculture and Industry Relative Income* (New York: Macmillan, 1956), p. 16.

11. Bellerby, *Agriculture and Industry Relative Income*, p. 187.

12. Bellerby, *Agriculture and Industry Relative Income*, p. 270.

13. These figures are not given by Bellerby for the postwar period but are estimated here by the procedures he outlined in *Agriculture and Industry Relative Income*, pp. 187–89.

14. Bushrod W. Allin, "Rural Influences on the American Politico-economic System," lecture to the U.S. Department of Agriculture Graduate School, April 1957.

15. Report to Congress from the *Commission on Increased Industrial Uses of Agricultural Products*, 85th Cong., 1st sess., June 1957, S. Doc. 45.

3. Technology, Surplus Disposal, and Supply Control

1. See *Journal of Farm Economics*, Proceedings Number, 41, no. 5 (December 1959).

2. See the outlook issues of the *Demand and Price Situation* (mimeographed) for the relevant data.

3. I was sorely tempted at this point to throw away the topic and outline of this paper and explore in detail our unconscious policy with respect to the development and dissemination of new knowledge in agriculture—to explore the large and unknown resource commitment in research and education that is propelling the technological revolution in agriculture. But this is clearly the subject of another paper, which is taking shape in my mind.

4. Food and Agriculture Organization of the United Nations, *Second World Food Survey* (Rome: FAO, November 1952).

5. See the discussion in Food and Agriculture Organization of the United Nations, *The State of Food and Agriculture, 1958* (Rome: FAO, 1958), pp. 9–33.

6. Ministry of Food, the Government of India, *Report on India's Food Crisis and Steps to Meet It*, April 1949, pp. 11–12.

7. Strictly speaking, adherence to the "additional principle" is limited to sales for foreign currency under PL 480, and there is continued pressure to engage in barter and concessional dollar sales that are competitive; but there is a grow-

ing general awareness that the leader of the free world cannot dump on its friends at will.

8. This emergency view of foreign surplus disposal of the present administration becomes abundantly clear in the recent testimony of Clarence L. Miller, assistant secretary of agriculture, on the "Food for Peace" bill, S. 1711. He states that "we do not believe that any greater rate of disposition would result under a five-year extension than can be attained under a one-year extension." But nowhere in his testimony does he give consideration to the needs and problems of the recipient countries in undertaking economic development. On the contrary, he voices the fear that authorizing supply commitments up to ten years would "provide for programming of certain commodities whether or not they are in surplus supply." For Miller, foreign surplus disposal is simply an avenue of eliminating visible surpluses that happen to be in the hands of the United States government.

9. This statement is not intended to constitute a rejection of multilateral commodity agreements; such agreements must play an important role in a controlled agriculture.

10. "Some Further Reflections on Supply Control," *Journal of Farm Economics*, November 1959.

4. Policy for the Twenty-first Century

1. I define a family farm as an organization in which the family makes all the important operating and investment decisions for the farm, owns a significant portion of the assets, and provides a significant amount of the labor required.

2. The classification of farms and estimates in this paragraph are from *Agricultural Outlook*, ERS, USDA, January–February, 1999.

3. Any measure of elasticity less than 1.0 is considered inelastic.

4. In this view of the global market for food products, global production in any one year is conceptualized as one great glob of products, a glob that is a function of many economic and physical variables: weather, national production and credit policies, past product price movements, past investment decisions both micro and macro, past research and development efforts both national and international, and on and on. In figure 1 that glob is represented by a vertical line (e.g., SS_{97}).

5. Two points need to be made here: further research may indicate that the cutoff figure of $250,000 is too low, excluding many legitimate commercial family farmers; and limited resource, residential, or retirement farms would not qualify for payments under this program.

5. The Export Solution

1. *The export demand* confronting Brazilian producers of soybeans can and will be higher than the global demand confronting soybean producers worldwide.

6. Saving the Family Farm

1. Although others may define the family farm differently, I believe my definition captures the essence of the concept: the family is the primary decision maker of the enterprise, the family has an important financial stake in the enterprise, and family members are active participants in the enterprise.
2. In the year or so between my writing "A Food and Agricultural Policy for the Twenty-first Century" (chapter 4 in this volume) and writing "Saving the Family Farm: The Case for Government Intervention" (chapter 6), I concluded from talking to a good number of farmers that my cutoff point for making cash subsidies to family farmers was too low. Thus, here I have raised it to $300,000 gross income. The concept remains unchanged.

7. American Agricultural Abundance

1. In this chapter I put in place a set of agricultural policies that have the capacity to achieve the goals Wendell Berry has set forth for farmers on one side and conservationists on the other. See "For the Love of the Land," *Sierra* 87, no. 3 (May–June 2002): 50–55.
2. For a description of these early food programs, see *Century of Progress* (Washington DC: USDA, 1963), pp. 183–88.
3. Actually we "sold" those commodities to the poor countries for their nonconvertible currencies. That made the bean counters in Congress happy and did not cost the recipient countries any of their scarce, precious foreign exchange.
4. The lively debate that took place over this issue is summarized in Vernon Ruttan, ed., *Why Food Aid?* (Baltimore: Johns Hopkins University Press, 1993), part 2, "Food Aid Controversies."
5. John Ikerd, paper presented at the National Conference of Block and Bridle, St. Louis MO, January 20, 2001.
6. This working definition was used to guide a comparative study of agriculture in four states: Montana, Minnesota, South Dakota, and Iowa. See Elizabeth Bird, Gordon Bultena, and John Gardner, eds., *Planting the Future: Developing an Agriculture That Sustains Land and Community* (Ames: Iowa State University Press, 1995), p. 57; research and publication underwritten by Northwest Area Foundation, St. Paul MN, publication coordinated by Center for Rural Affairs, Walthill NE.

7. Supporters of this policy will benefit from a careful reading of a new book, Dana L. Jackson and Laura L. Jackson, eds., *The Farm as Natural Habitat: Reconnecting Food Systems with Ecosystems* (Washington DC: Island Press, 2002). Program administrators and fieldworkers responsible for developing individual farm plans designed to convert those farms to sustainable operations should read chapter 14, "Composing a Landscape."

8. See Bird, Bultena, and Gardner, *Planting the Future,* chap. 6.

9. "Failing in Style," *Choices: The Magazine of Food, Farm and Resources* (AAEA, Ames IA), no. 1 (2001).

10. Deborah E. Popper and Frank Popper, "The Great Plains: From Dust to Dust," *Planning*, December 1987, pp. 12–18.

11. For a discussion of this concept, see Daniel Licht, *Ecology and Economics of the Great Plains* (Lincoln: University of Nebraska Press, 1997), pp. 132–76.

12. Walter Prescott Webb, *The Great Plains* (New York: Houghton Mifflin, 1936), 420–23.

13. Florence Williams, "Plains Sense: Frank and Deborah Popper's Buffalo Commons Is Creeping toward Reality," *High Country News* 33, no. 1 (January 15, 2001).

14. A statement issued by the NRCS of the USDA, January 2001, reads as follows: "Excessive erosion continues to be a serious problem in many parts of the country. . . . Excessive erosion of 1.3 billion tons per year leads to concerns about sediments, nutrients, and pesticides impacting water quality, as well as air quality in wind erosion areas of the West, Midwest, Northern Plains, and Southern Plains."

15. Gyles W. Randall, "Present-Day Agriculture in Southern Minnesota—Is It Sustainable?" Southern Research and Outreach Center, University of Minnesota, Waseca, 56093–4521.

acreage, farm: federal land leasing and, 17–18; frontier settlement and, 24, 125–26; future farm policy and, 104–5; planting allotments, 108; production control and, 16, 65–68, 72–73. *See also* allotments, acreage

advertising, 86–87

aggregate demand, 10, 11, 19–24, 35–37, 43, 54, 60–61, 62, 66–67, 74, 76, 88–89. *See also* demand, farm product

aggregate supply, 9, 10, 14, 19, 24–26, 35–37, 40–41, 43, 45, 58–59, 60–61, 66–67, 76. *See also* supply, farm product

agribusiness: agriculture and, 18, 65; family farms and, 68–69, 94–95, 101–3; farm programs and, x; growth in American agriculture, 2; market organization and, 29–33; production contracts, 69; production control and, 111–13; restructuring agriculture and, 127–29; technology adoption and, 25–27

agriculture: as a business, 18, 95; conventional vs. sustainable, 135–42; developing nations and, 51–55; family farms and, 91–98; future farm policy and, 77–82; future of commercial, 107; market organization of, 29–33; monopoly in, 71–72, 79, 85–86, 102–4; national economy and, 15; restructuring of U.S., 113–29, 132, 147 n.14; technology adoption and, 33–39, 41–43, 102–3. *See also* sustainable agriculture

aid, foreign, 45–46, 49–55

allotments, acreage, 108. *See also* acreage, farm

alternatives, production, 13–14

American Farm Economics Association, 3, 45

animals. *See* livestock; wildlife

assets, farm: farm output and, 47; federal refinancing of, 120; sustainable agriculture and, 140–41; technology adoption and, 35–39. *See also* investment

biotechnology: family farms and, 101–3; future farm policy and, 79, 103–4; genetically modified organism, 72; production contracts, 69. *See also* technology

birth rate and population, 23

breeding, plant and animal, 10, 25, 48, 72

"Buffalo Commons," 122, 132–33

Bureau of Agricultural Economics (BAE), xi, 1

capital. *See* assets, farm; investment

Carson, Rachel, x

climate: future farm policy and, 82, 100; global markets and, 85–86, 97, 131; supply vs. demand and, 76; sustainable agriculture and, 137–38; world hunger and, 98–99

Commodity Credit Corporation, 99

commodity markets: domestic, 3–4; farm prices and, 14, 96–98; future farm policy and, 72–73; global, 4; market organization and, 29–33; production control and, 16. *See also* export markets

Conservation Reserve Program (CRP), 81

consumers: food demand by, 84–85; food price and, 22; nutrition and, 12, 37–38; subsidies and low-income, 13. *See also* income; population growth

consumption, food: farm prices and, 11; future and, 28–29; nutrition and, 12–13;

consumption (*continued*)
  population growth and, 20–24; world
  population and, 50–52. *See also* food
contracts, production: future farm policy
  and, 79, 102–3; monopoly and, 71–72;
  technology adoption and, 69. *See also*
  industrialization, farm
control, production. *See* production control;
  supply control
cooperatives, marketing, 117–19. *See also*
  marketing
CRP. *See* Conservation Reserve Program
  (CRP)

dairy. *See* livestock; milk
deficiency payments, 66–67, 108
demand, farm product: components of, 62;
  export markets and, 73–77, 83–90; farm
  policy and, 66–67; farm prices and, 19,
  45–46, 96–98; foreign economic development and, 58–60; future and, 27–29, 131;
  technology adoption and, 35–39. *See also*
  aggregate demand; supply, farm product
depression: agriculture and business, 15, 66–
  67, 107, 131; farm price, 3, 19; farm supply
  vs. demand and, 28–29, 111; income elasticity and, 22–23
developing countries, 45
diet. *See* food; nutrition
disease control, crop, 10, 25, 48
diversions, acreage. *See* acreage, farm

economic aid, 45–46, 49–55
economic development, 52–55
Economics, Bureau of Agricultural, xi, 1
economy, U.S.: agriculture and the, 15;
  export markets and, 83–90; farm depression and, 3; farm supply vs. demand and,
  28–29; future farm policy and, 5, 18; sustainable agriculture and, 116; technology
  adoption and, 41–43. *See also* government;
  technology
education: foreign economic development
  and, 54–56; future farm policy and, 98,
  103; research and, 48, 144 n.3; restructuring agriculture and, 123; rural, 69–70; sustainable agriculture and, 116–17; technology adoption and, 26–27, 33–34
Eisenhower Administration, ix

emigration. *See* labor, farm; migration, off-
  farm
environment: agriculture and the, x, 107, 147
  n.14; Conservation Reserve Program and,
  81; ecological reserves and the, 122, 124;
  restructuring agriculture and the, 126–29;
  rural quality of life and, 70–71, 132–33;
  sustainable agriculture and, 116, 137–39
Environmental Protection Agency, 80
export markets: farm prices and, 4–5, 65, 83–
  90, 96–98; farm programs and, x. future
  farm policy and, 72–82, 104–5; future of,
  131; global supply and, 145 n.4; subsidies
  and, 111–13; surplus disposal and, 45–46,
  49–55, 111. *See also* commodity markets;
  markets

factory farms. *See* industrialization, farm
FAIR. *See* Federal Agricultural Improvement
  and Reform Act (FAIR)
family farms: biotechnology and, 72; decline
  of, 2, 67–70; defined, 91–94, 145 n.1; farm
  policy and, 65; farm prices and, 14–15;
  future farm policy and, 79, 103–4, 131–32;
  industrialization and, 71–73; production
  control and, 17–18; technology adoption
  and, 39, 101–3
famine. *See* hunger
Farm Credit Administration, 1
farm depression. *See* depression; farm price
farm income. *See* income
farm output. *See* supply, farm product
farm policy. *See* policy, farm
farm prices: commodity markets and, 4–5;
  demand for food and, 11–12, 43, 45–46;
  falling, 9–10, 36, 65–66; family farms and,
  94–98; frontier settlement and, 24; future
  farm policy and, 27–29, 77–82, 98–101;
  globalization and, 71, 83–90; government
  programs and, 108–11; market organization and, 29–33, 41–42; nonfarm food services and, 21–22; production alternatives
  and, 13–14; restructuring agriculture and,
  133; society support of low, 34–37; supply
  vs. demand and, 19, 35–39, 58–62; surpluses and, 2–3; sustainable agriculture
  and, 118–20, 135–42
farm size: family farms and, 67–69, 91–94,
  101; frontier settlement and, 24, 125–26;

sustainable agriculture and, 141; technology adoption and, 37, 39. *See also* acreage,
farm
Federal Agricultural Improvement and
Reform Act (FAIR), 105, 111, 115, 120
Federal Farm Board, 99
federal land purchase and leasing, 17–18
fertility, crop, 10, 137–38, 140
food: demand for, 84–85; foreign economic
development and, 52–55; future farm policy and, 77–82, 103; government programs
and, 108–11; inelastic demand for, 10–12,
41–43, 73–77, 96–98; politics, 45; population growth and, 20–24; safety and farm
programs, x, 77; supply vs. demand and,
28–29, 66–67, 73–77; technology adoption
and, 37–38; world hunger, 50–52; world
poverty and, x. *See also* consumption,
food
Food and Agriculture Organization (FAO),
50, 52, 55–56. *See also* United Nations
Food and Drug Administration, 77
Food for Peace, 110, 144 n.7, 145 n.8
Ford Foundation, 2, 51
foreign aid, 45–46, 49–55
foreign markets: exports and, 83–90; farm
income and, 15; farm prices and, 4–5; surplus disposal and, 52–55
Freeman, Orville, 1, 45
free trade. *See* marketing; trade agreements
frontier settlement, 24, 125–26
future: agriculture and the, 107; farm policy
and the, 65, 70–71, 77–82; restructuring
agriculture and the, 123–29, 131–33; supply
vs. demand and the, 27–29

genetically modified organism (GMO), 72
global markets. *See* export markets; markets
global warming, 131. *See also* climate
government: farm policy and, 36–37, 65–67;
food programs, 108–11; free trade and, 87–
88; future farm policy and, 70–73, 77–82;
98–101, 103–4; global markets and, 85–86,
96–98; Homestead Act of 1909, 121–22;
rural programs, 70–71; subsidies and the,
111–13; surplus disposal and, 52–55. *See
also* economy, U.S; politics; society
grain reserves. *See* reserves; storage
grazing, livestock. *See* livestock

growth: birth rate and population, 23; farm
output and population, 10; food and
population, 20. *See also* population
growth

Harvard University, 1
health services, rural, 69–70
Homestead Act of 1909, 121–22
Humphrey, Hubert, 1
hunger: future farm policy and, 77–78, 98–
99; surplus disposal and, 45, 49–55, 108–
11. *See also* poverty, world

immigration. *See* labor, farm; migration, off-
farm
income: demand for food and, 11–12, 76, 85–
86; elasticity and food, 20–24, 45; equity
of farm and urban, 15–16; family farms
and farm, 91–94; farm prices and, 14–15,
60; globalization and farm, 71; government programs and farm, 68, 72–73;
incentive, 39–41; market organization and
farm, 29–31; mechanization and impact
on, 32–33; nutrition and, 12–13; supply vs.
demand and farm, 19, 28–29, 35–39, 66;
surpluses and, 2–3; technology adoption
and, 38–39
industrialization, farm: agriculture and, 71–
73; family farms and, 91–94, 101–3; future
farm policy and, 79–80. *See also* contracts,
production
insect control, crop, 10
Institute of Agriculture and Trade Policy, xi–
xii, 3, 65
investment: family farms and, 69; foreign
economic development and, 54–58; future
farm policy and, 77–82; global markets
and, 85–86, 96–98; supply vs. demand
and farm, 76. *See also* assets, farm
irrigation, 10, 25, 48, 76, 127

Kennedy, John F., ix, 1, 45, 110

labor, farm: family farms and, 91–94; frontier settlement and, 24; future farm policy
and, 80–81; restructuring agriculture and,
123; supply control and, 60–62; technology adoption and, 38–39, 47–49. *See also*
migration, off-farm; population

land: Conservation Reserve Program and, 81; farm output and, 47; farm prices and, 100–101; production control and, 17–18, 60–62, 107–11; restructuring agriculture and, 123–29

leasing, federal land, 17–18

livestock: breeding, 10, 25; factory farms and, 71, 80, 101–2; marketing cooperatives and, 118; production control and, 17; restructuring agriculture and, 113–29, 132–33; surplus disposal and, 46; sustainable agriculture and, 135–42. *See also* milk; wildlife

marketing: agreements, 16; associations, 60–61; free trade and, 111–13; orders, 13, 16; quotas, 16, 60–62, 107; sales promotion and, 86–87; sustainable agriculture and, 117–19, 133. *See also* contracts, production; cooperatives, marketing; trade agreements

markets: commodity, 3–4; farm prices and, 29–33; technology adoption and, 41–42. *See also* export markets

mechanization: animal power and, 24–25; human labor and, 32–33, 36, 39, 47–48; impact on farm prices, 10. *See also* technology

merchandising, food, 21

Metropolitan Cooperative Milk Producer's Bargaining Agency, 2–3, 9

migration, off-farm: depression and, 29; mechanization and, 25; production control and, 17; restructuring agriculture and, 123; supply vs. demand and, 59–60; technology adoption and, 38–40. *See also* labor, farm; population

milk: demand for food and, 11–12; Metropolitan Cooperative Milk Producer's Bargaining Agency, 2–3; production control and, 16–17; sustainable agriculture and, 135–42. *See also* livestock

monopoly in agriculture: food production and, 71–72, 102–3; future farm policy and, 79, 103–4; global markets and, 85–86. *See also* industrialization, farm

Montana State University, 1

National Labor Relations Board, 81

national sustainable farm program: "Buffalo Commons" and the, 122; marketing coop-

eratives and the, 118–20; proposal for a, 123–25. *See also* sustainable agriculture

New York, 2–3, 9

1963 Wheat Referendum, ix

Northwest Area Foundation, xii, 118–19

nutrition: animal, 25; food consumption and, 12–13; future farm policy and, 103; government programs and, 108–11; technology adoption and, 37–38

Occupational Safety and Health Administration, 80–81

output, farm. *See* supply, farm product

packaging, food, 21

parity and price supports, 14

patent rights, 71, 102–3

Pennsylvania State University, 1

PL 480, 110, 144 n.7, 145 n.8

plant breeding, 25

policy, farm: foreign economic development and, 58–62; future, 5, 65, 70–73, 77–82, 98–101, 103–4; goals in America, 65–70; restructuring agriculture and, 113–29; surpluses and, 52–55, 58, 107–11; technology adoption and, 49. *See also* subsidies

politics: farm policy and, 1, 58, 67, 81–82, 103–4; food, 45; food programs and, 108–11; free trade and, 87–88; restructuring agriculture and, 125–29; rural life and, 69–70; subsidies and, 111–13, 122; surpluses and, 50. *See also* government

Popper, Deborah E. and Frank, 122

population: declining rural, 69–70, 126–27; demand for food and, 11–12. *See also* labor, farm; migration

population growth: farm output and, 10, 45–46; food consumption and, 20, 22–23; global markets and, 85–86; off-farm migration and, 38–39; supply vs. demand and, 76, 96–98; technology adoption and, 27–28, 37–38; world hunger and, 50–52. *See also* growth

poverty: rural, 69–70; world, x, 41, 58, 110–11. *See also* hunger

Powell, John Wesley, 125–26

prices, farm. *See* farm prices

prices, food. *See* food

price supports: farm prices and, 66–68; future farm policy and, 72–73, 104–5; global markets and, 65; parity and, 14; supply vs. demand and farm, 28–29; surpluses and, 13, 45–46, 107–11

processing, food, 21

production, farm: alternative and price supports, 13–14; developing nations and, 51–52; export markets and, 83–90; family farms and, 91–94, 103–4; future and, 27–29; income and, 39–41; market organization and, 29–33; sustainable agriculture and, 99–100, 140; technology adoption and, 25–27, 33–37, 101–3

production control: foreign aid and, 45; free trade and, 111–13; future farm policy and, 72–73, 104–5; methods of, 16–18; price supports and, 36–37; surpluses and the use of, 9, 107–13. *See also* supply control

profitability of sustainable agriculture, 135–42

public assistance and food, 108–11

purchases, federal land, 17–18

referendum, 1963 Wheat, ix

research: agriculture and impact of, 48–49; future farm policy and, 77, 103; global markets and, 85–86, 96–98; society support of, 33–37, 41–43; supply vs. demand and, 76; sustainable agriculture and, 116; technology and, 25, 144 n.3

reserves: grain, 98–99, 120; wildlife ecological, 122, 124. *See also* storage

Roosevelt, Franklin D., 70, 81

Roosevelt, Theodore, 70

Rural Electrification Administration (REA), 70

safety, worker, 80–81. *See also* food, safety

sales. *See* marketing

Saudia Arabia, 2

services, nonfarm, 21–22

Silent Spring (Carson), x

society: farm policy and, 36–37, 41–43, 70–73, 77–82, 98–101; support of technological advances by, 33–35; sustainable agriculture and, 114. *See also* government

storage: food prices and, 21; grain reserves and, 78, 120; surplus, ix, 45–46. *See also* reserves

subsidies: food consumption and, 13–14; future farm policy and, 79, 104–5, 145 n.5, 146 n.2; global markets and, 85–86; government programs and, 68, 111–13, 139; restructuring agriculture and, 127–29, 131–33; sustainable agriculture and, 99–100, 115–16, 120. *See also* policy, farm; price supports

supply, farm product: export markets and, 83–90; farm policy and, 65–67; farm prices and, 19, 45–46, 96–98; food and, 24; future and, 27–29, 131; society support of, 34–37; surplus disposal and, 52–55; technology adoption and, 35–39, 143 n.8. *See also* aggregate supply; demand, farm product

supply control: foreign economic development and, 58–62; global markets and, 5; Kennedy administration, ix; methods of, 16–18, 60–62. *See also* production control

surpluses: American agriculture and, 2–3; disposal of, 45–46, 52–55, 108–11, 144 n.7, 145 n.8; export markets and, 49–50, 83–90; farm prices and, 9, 13–14, 111–13; future farm policy and, 72–73; storage of, ix, 78; supply vs. demand and farm, 28–29, 45–46; world poverty and, x

sustainable agriculture: case studies in, 135–42; defined, 113–15; economics of, 118–19; environment and, x; future farm policy and, 79–81, 98–101; national program proposal for, 115–29; restructuring agriculture and, 132–33, 147 n.7. *See also* agriculture; national sustainable farm program

technology: agriculture and impact of, 2–3, 5, 33–43; family farms and, 68, 94, 101–3; farm assets and impact of, 35–37; farm market organization and, 29–33; farm output and, 24–26, 46–49, 108, 143 n.8; farm prices and, 9–10, 15; future farm policy and, 103–4; global markets and, 85–86, 96–98; human labor and, 38–39, 47–49; income and, 39–41; population growth

technology (*continued*)
and, 27–28; societal support of, 41–43;
supply vs. demand and, 76; sustainable
agriculture and, 116. *See also* biotechnology; economy, U.S; mechanization
trade agreements, 87–90. *See also* marketing
transportation, food, 21
treadmill theory, 19, 31–32, 41–43

unit costs, 29–33, 35, 42, 48
United Nations, 1–2, 50, 55. *See also* Food
and Agriculture Organization (FAO)

University of California, 1
University of Minnesota, xii, 1, 3,
83

War Food Administration, 1
weather. *See* climate
wheat referendum, 1963, ix
wildlife, 81, 121, 124. *See also* livestock
worker protection, 80–81
world poverty. *See* hunger; poverty,
world

In the Our Sustainable Future series

Volume 1
Ogallala: Water for a Dry Land
John Opie

Volume 2
Building Soils for Better Crops: Organic Matter Management
Fred Magdoff

Volume 3
Agricultural Research Alternatives
William Lockeretz and Molly D. Anderson

Volume 4
Crop Improvement for Sustainable Agriculture
Edited by M. Brett Callaway and Charles A. Francis

Volume 5
Future Harvest: Pesticide-Free Farming
Jim Bender

Volume 6
A Conspiracy of Optimism:
Management of the National Forests since World War Two
Paul W. Hirt

Volume 7
Green Plans: Greenprint for Sustainability
Huey D. Johnson

Volume 8
Making Nature, Shaping Culture: Plant Biodiversity in Global Context
Lawrence Busch, William B. Lacy, Jeffrey Burkhardt, Douglas Hemken,
Jubel Moraga-Rojel, Timothy Koponen, and José de Souza Silva

Volume 9
Economic Thresholds for Integrated Pest Management
Edited by Leon G. Higley and Larry P. Pedigo

Volume 10
Ecology and Economics of the Great Plains
Daniel S. Licht

Volume 11
Uphill against Water: The Great Dakota Water War
Peter Carrels

Volume 12
Changing the Way America Farms: Knowledge and Community in the Sustainable Agriculture Movement
Neva Hassanein

Volume 13
Ogallala: Water for a Dry Land, second edition
John Opie

Volume 14
Willard Cochrane and the American Family Farm
Richard A. Levins

Volume 15
Raising a Stink: The Struggle over Factory Hog Farms in Nebraska
Carolyn Johnsen

Volume 16
The Curse of American Agricultural Abundance: A Sustainable Solution
Willard W. Cochrane